The
MOON
BOOK

by Kim Long

Johnson Books: *Boulder*

© 1988 by Kim Long
Cover design, book design, and illustrations by Kim Long
First Edition
1 2 3 4 5 6 7 8 9
ISBN 1-55566-028-2
LCCCN: 88-81081

Printed in the United States of America by
Johnson Publishing Company
1880 South 57th Court
Boulder, Colorado 80301

The Moon Book is a companion text to *The Moon Calendar*, also by Kim Long. *The Moon Calendar* is an annual publication in poster format (31½" x 20½") printed in nocturnal black and lunar white, and packaged in clear, heavy plastic tubes with vinyl caps. The phases of the moon are graphically displayed for every day of the year, with information about phases, lunar eclipses, and dates of apogee and perigee. *The Moon Calendar* is also available as a gift or display card (6½" x 10½"). *The Moon Calendar* for each new year is available in August of the preceding year.

For information about ordering *The Moon Calendar* or inquiries about quantity or trade discounts, contact the publisher:

Johnson Books
1880 South 57th Court
Boulder, CO 80301
303-443-1576

Special thanks to . . .
Cathie Havens, S & S Optika (Englewood, Colorado)
Larry Sessions, Staff Astronomer, Denver Museum of Natural
    History (Denver, Colorado)
Michael McNierney, Johnson Books (Boulder, Colorado)
Kathleen Cain, Front Range Community College (Westminster,
    Colorado)
Denver Public Library
Hal Stephens
Mary Hagen, Lunar and Planetary Institute (Houston, Texas)
David Black, Defense Mapping Agency (St. Louis, Missouri)
U.S. Geological Survey
U.S. Government Bookstores
Celestron Telescopes
Marie Lucas, U.S. Naval Observatory (Washington, D.C.)

Other publications by the author:
The Moon Calendar (annually since 1982, Johnson Books,
    Boulder, Colorado).
The American Forecaster (annually since 1984, Running Press,
    Philadelphia).
The Trout Almanac (1987, Johnson Books, Boulder,
    Colorado).
The Daily Planet Vacation Almanac (with Terry Reim, 1986,
    World Almanac Publications, New York).
Kicking the Bucket (with Terry Reim, 1985, Wm. Morrow &
    Company, New York).

Table of Contents

Introduction

The moon has long been a fascinating subject for artists and illustrators. This author was once an illustrator himself, with the moon an important focus for a project—an annual almanac—that required a picture of the phase of the moon for every day of the year. There were no sources of photographs that showed a complete, day-by-day cycle of the lunar month, and the illustrator was stuck with the problem of creating one.

The resulting daily illustrations of the moon were used every year for more than a decade in this almanac project. The artist, during an idle moment in the midst of this period, found another use for the hundreds of little moon drawings—a moon calendar featuring every phase laid out side by side in poster format.

This moon calendar now has a successful life of its own and grows more popular every year. A by-product of this popularity, however, has been an increasing number of questions from calendar fans. A person who creates a lunar calendar, it seems, is also expected to have the answer to the mysteries of the lunar cycle.

This book project was developed to provide a handy source of answers to common, and not-so-common, questions about the moon. It is hoped, however, that restless minds will not be satisfied with simple answers, but be inspired to search for more than any one book can provide. Meanwhile, the former illustrator is still searching for more uses for the hundreds of little moon drawings: lunar wallpaper, lunar underwear, lunar ties, ... the list grows looney.

The diameter of
the moon is less than the
width of the United States.

EARTH

MOON

The earth and the moon at the same scale

(Photograph courtesy of NASA; moon photo courtesy of Lick Observatory)

Moons of Jupiter

The moon and other solar system

Moons of Saturn

bodies at the same scale.

(Photograph courtesy of NASA)

The Vital Statistics of the Moon

Mean distance of the moon from earth
238,857 miles (382,176 km)
60.27 earth radii
0.002 570 a.u.

Greatest distance of the moon from earth
252,710 miles (404,336 km)

Shortest distance of the moon from earth
221,463 miles (354,340 km)

Circumference
6,790 miles (10,930 km)
0.27 of earth's circumference

Diameter
2,160 miles (3,476 km)
0.27 of earth's diameter

Mean radius
1,080 miles (1,738 km)

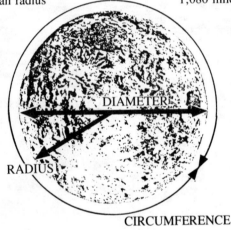

CIRCUMFERENCE

Mean angular diameter   31' 07"

Mass 8×10^{19} tons (7.35×10^{22} kg)
0.0123 earth's mass

Mass ratio (earth/moon)   81.301

Volume   (2×10^{25} cm^3)
0.0204 earth's volume

Mean density 3.34 g/cm^3

(3.34 more dense than water)

0.6 earth's density

Gravity at the surface 5.31 ft/sec^2 (1.62 m/s^2)

0.1667 g (1/6 earth's gravity)

Escape velocity   1.48 miles/sec (2.38 km/sec)

Mean inclination to lunar equator   6° 41'

Mean orbital inclination to ecliptic   5° 08' 43"

Inclination of equator to orbit   1° 32'

Period of revolution of perigee   3,232 days

Orbital direction   east (counterclockwise)

Mean orbital speed   2,287 miles/hour (3,680 km/hr)

33 minutes arc/hour

Mean centripetal acceleration $0.002\ 72$ m/s^2

0.0003 g

Mean eccentricity of orbit   0.0549

(mean eccentricity of earth's orbit is 0.0167)

Regression of nodes   18.5995 years (19.538 degrees per year)

Rotation period   27 days, 7 hours, 43 minutes, 11.5 seconds

Surface temperature   273° F (120° C) day

-244° F (1153° C) night

Visible surface   41 percent at any one time

18 percent additional surface visible due to librations

59 percent total visible surface

Parallax   0.9507 degrees

Moon's angular diameter   0.5181 degrees

Magnitude of full moon   -12.5

Average albedo   0.12

See Glossary (page 116) for explanations of unfamiliar terms.

km: kilometer	g: gravity
kg: kilogram	m: meter
a.u.: astronomical unit	s or sec: second
cm: centimeter	hr: hour

Synodic month (new moon to new moon)   29.53059 days
                        29 days, 12 hours, 44 minutes, 2.8 seconds
Sidereal month (star to star)   27.32166 days
                        27 days, 7 hours, 43 minutes, 11.5 seconds
Anomalistic month
(apogee to apogee or perigee to perigee)   27.55455 days
                        27 days, 13 hours, 18 minutes, 33.2 seconds
Nodical month, draconic month (node to node)   27.21222 days
                        27 days, 5 hours, 5 minutes, 35.8 minutes
Tropical month
(first point of Aries to first point of Aries)   27.321582 days
                        27 days, 7 hours, 43 minutes, 4.7 seconds

The earth and the moon orbit in a counterclockwise direction.

The moon's orbital distance
at apogee is equivalent to
32 earth diameters.

The moon's orbital distance
at perigee is
equivalent to
28 earth diameters.

The moon is slightly egg-
shaped, not round. The large
end of this "egg" points toward
the earth. At most, this bulge
adds only about 3 miles to the
radius of the moon.

Moon Speed

The moon rotates around its axis once every 29 1/2 days. This is the same time it takes for the moon to complete one revolution around the earth. The similarity of these two figures is no coincidence. The moon rotated much faster in the past, but one of the effects of the earth's gravity on the moon over millions of years has been to slow down the rotation until the moon has become "locked in step" with the earth.

The earth rotates at about 1,000 miles an hour, as measured from a point on the surface at the equator. By comparison, the moon rotates at only 10 miles an hour.

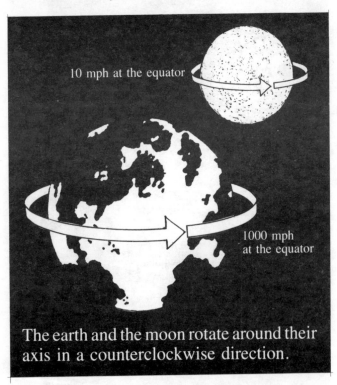

10 mph at the equator

1000 mph at the equator

The earth and the moon rotate around their axis in a counterclockwise direction.

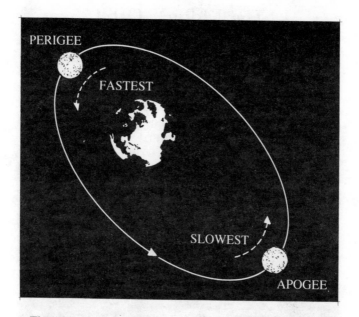

The orbital speed of the moon is much faster than its rotation. The average speed of the moon on its monthly trip around the earth is 2,287 miles an hour (3,680 kilometers an hour). However, the moon's orbit is not a circle but an ellipse. The effect of an elliptical path on the speed of an orbiting object is to change the speed at different parts of the orbit.

When the moon is closest to the earth, it is traveling at its maximum speed, 2,429 miles an hour. At the farthest point from the earth, the speed is slowest, 2,153 miles an hour. An observer on the surface of the earth sees this orbital speed as the movement of the moon across the sky at about one full moon's width per hour. The variations in speed from fastest to slowest are not very noticable from this vantage point.

The Moon's Orbit

The moon orbits around the earth in an elliptical path. An ellipse is an elongated circle; the degree to which it is elongated is called its eccentricity. The eccentricity of the moon's orbit is very slight, too small to accurately depict in an illustration that would fit in this book. The difference between the shortest and longest distances from the earth to the moon, in fact, is only about the width of four earths *(see page 15)*.

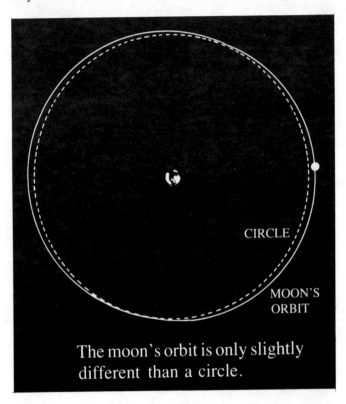

CIRCLE

MOON'S ORBIT

The moon's orbit is only slightly different than a circle.

The orbit of the moon, however, is more complex than just that of one body moving around another. The moon and the earth are actually orbiting around a common center of gravity called a barycenter that is created by the combination of the two masses. The barycenter of the earth-moon system is itself constantly moving as both bodies are in motion, one orbiting around the other, and both rotating around their axis.

The earth, being much larger than the moon, "pulls" the barycenter much closer to itself than to the moon. The barycenter located below the surface of the earth, about 3/4 of the distance from the center of the planet.

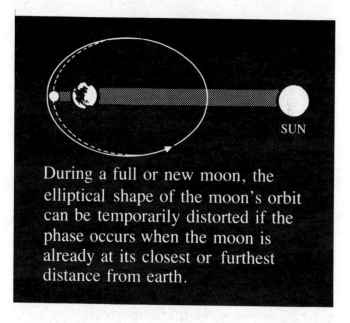

SUN

During a full or new moon, the elliptical shape of the moon's orbit can be temporarily distorted if the phase occurs when the moon is already at its closest or furthest distance from earth.

The Rotation of the Moon

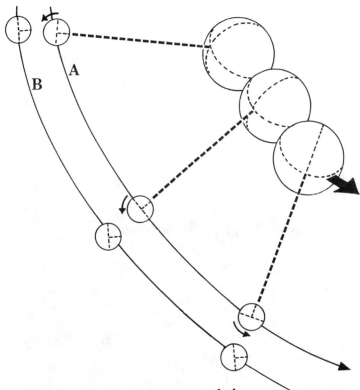

As the moon orbits around the
earth, it is also slowly rotating,
always keeping the same side facing the
earth (A). If the moon were not rotating, the
side facing earth would change (B).

Lunar Month

The moon completes one orbit around the earth every 29.5 days. This period is called a lunation, lunar month, or synodic month. A lunation begins at the time of new moon, and one lunation is officially 29 days, 12 hours, 44 minutes, 2.8 seconds long. If the lunar cycle is measured by timing it in relationship to the position of a specific star (this is called a sidereal cycle), the lunar month is only 27 days, 7 hours, 43 minutes, 11.5 seconds long. The difference between the two measurements is about 2 days a month, which is about how much a full moon will "lag" behind the calendar from one month to the next.

A lunation is a very visible cycle that is easy to observe. Most people, however, notice the full moon instead of the new moon as it is more obvious and think of the lunar cycle as running from full moon to full moon. Either method produces the same measurement, with the lunar cycle running like clockwork, never fast or slow.

Lunations are numbered in sequence. The sequence began with Lunation Number 1, on January 16, 1923. There are 13 lunations in every calendar year because calendar months are longer than lunar months (with the exception of February).

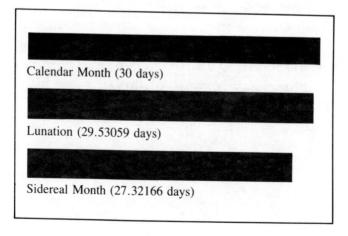

Calendar Month (30 days)

Lunation (29.53059 days)

Sidereal Month (27.32166 days)

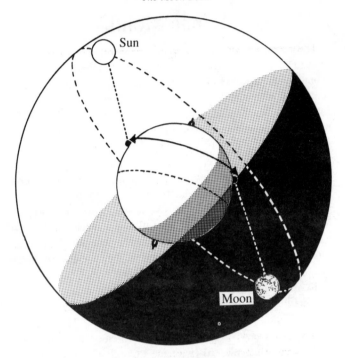

The earth's axis is tilted to the path of the sun and the moon, both following almost the same line across the sky. An observer will rotate into a different perspective from night to day because of this tilt, making the moon appear high in the sky when the sun is low, and vice versa.

PLANE OF THE
EARTH'S ORBIT

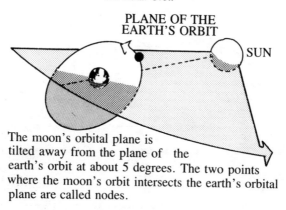

SUN

The moon's orbital plane is
tilted away from the plane of the
earth's orbit at about 5 degrees. The two points
where the moon's orbit intersects the earth's orbital
plane are called nodes.

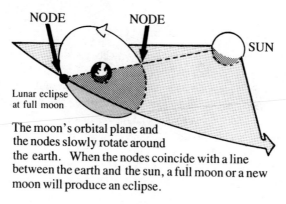

NODE   NODE

SUN

Lunar eclipse
at full moon

The moon's orbital plane and
the nodes slowly rotate around
the earth. When the nodes coincide with a line
between the earth and the sun, a full moon or a new
moon will produce an eclipse.

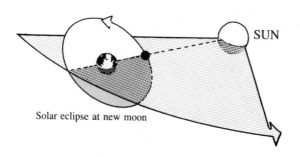

SUN

Solar eclipse at new moon

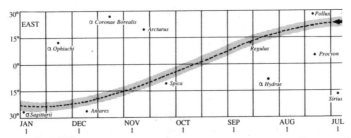

The sun always follows the same path
across the sky, referred to as the ecliptic.

The Ecliptic

The sun's path across the sky—actually caused by the movement of the earth—is called the ecliptic. The ecliptic is a fixed path relative to the stars, making an unchanging map of the sky that can be used to find directions and tell time. Prominent stars and constellations are like roadsigns on this path, which also includes the twelve traditional constellations of the zodiac. The moon's path is tilted to the ecliptic by about 5 degrees, and since this tilt is itself rotating, the lunar path over time will eventually "sweep out" an area that is 5 degrees above and below the path of the sun. Since the moon is moving almost twelve times faster than the sun around this stellar track, it makes a complete circuit in less than a month, a journey that takes the sun one year to complete.

Moondogs, Moonbows, and Moon Pillars

A strong source of light such as the sun or moon sometimes produces interesting optical effects when combined with the right atmospheric conditions. A rainbow—produced when sun-

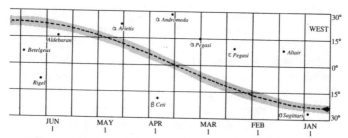

The moon's path is tilted to the ecliptic at an angle of about 5° (shaded area).

light is refracted through water droplets in the air—is perhaps the most familiar of these effects. When the light from the moon is refracted through water droplets, a similar effect can be produced. This prismatic display is known as a moonbow or lunar rainbow, but is much less intense than a solar rainbow. Moonbows are most likely to be seen when the moon is full.

Moondogs, also known as mock moons, are also produced by the interaction of moonlight and moisture in the atmosphere. With the right combination of humidity and angle to an observer, paler images of the moon can sometimes be seen off to the side of the central image. The official name of this halo effect is *paraselene* if the extra image is seen 22 or 46 degrees away, *parantiselene* if the image is at 90, 120, or 140 degrees, and *antiselene* if it is at 180 degrees.

Moon pillars are another form of halo related to the moon, but are rarely visible. Moon pillars can be seen when the moon is near to the horizon, and appear as shafts of light above and below the moon.

Lunar Phases

The moon appears to change shape as it moves through its monthly cycle. The moon itself is not changing shape, but the part that is illuminated by the sun does change, giving us the distinctive lunar phases. The four official phases that are included in calendars and almanacs are new moon, first quarter moon, full moon, and last quarter moon.

The new moon is actually invisible most of the time, as it is between the sun and the earth, and not even a thin crescent is illuminated by the sun. People can sometimes observe the new moon, however, when it being eclipsed and is illuminated by the secondary reflected light from the earth (earthshine). Earthshine is very weak compared to sunlight, however, and as the new moon is above the horizon during daylight hours, the bright light of daytime is very effective in cancelling the earthshine effect.

The first and last quarter moons mark the halfway points between the new moon and full moon. The first quarter moon is always "first" and is distinguished by the illuminated half of the lunar surface being on the right-hand side. The last quarter moon is the reverse, with the illuminated half on the left-hand side.

The lighted part of the moon always points the way to the sun. If the lighted half is on the right, the sun is on the right (west), meaning the sun is ahead of the moon. If the left half is lighted, the moon is ahead of the sun, and the sun is on the left (east).

The sequence of the lunar phases always proceeds with the lighted part of the moon growing from right to left until the full moon, then receding from right to left until the new moon. You can always tell where the moon is in its cycle if you remember that the changing pattern of light and dark always goes from right to left.

The new moon is also called the "dark of the moon," as that is when it is totally dark. The period of darkness officially lasts for only a second—that point when the moon is directly

between the sun and the earth. However, observers on earth don't see any "slivers" of light at the edges of the moon just before and after this point because the new moon is too close to the sun to see anything clearly, and the sperical shape of the moon prevents the light from being visible until it has moved some distance across the surface.

During the period around the new moon, the moon follows the sun very closely, starting to lag behind the sun after one or two days. The first visible crescent moon (with the crescent on the right) is usually spotted two or three days after the new moon. Observers can see this young crescent moon, as it is sometimes called, just after sunset, with the moon following the sun down over the western horizon. The earliest that an observer on earth has ever seen the young crescent moon is about 14 hours after new moon.

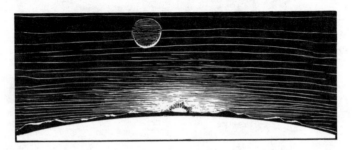

As the moon "grows," or waxes, the lighted portion on the right gradually increases until it forms an almost perfect half circle, the first quarter moon. The quarter moons are created by the position of the moon in relation to the earth and sun. At this point in the cycle, the orbit of the moon has moved it to a position off to the side of the line connecting the earth and the sun. At the point of first quarter moon, the moon is 90 degrees (perpendicular) to this imaginary line. Technically, this phase is called a quadrature.

The edge of light that gradually creeps across the lunar surface is called the terminator. The terminator is rarely an even line, as the rough surface and uneven curvature of the moon distort it, a phenomenon visible through binoculars or telescopes.

As the moon continues waxing, it continues to fall farther and farther behind the movement of the sun. The period after

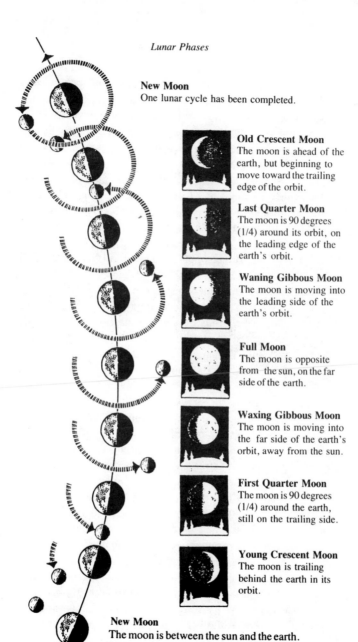

New Moon
One lunar cycle has been completed.

Old Crescent Moon
The moon is ahead of the
earth, but beginning to
move toward the trailing
edge of the orbit.

Last Quarter Moon
The moon is 90 degrees
(1/4) around its orbit, on
the leading edge of the
earth's orbit.

Waning Gibbous Moon
The moon is moving into
the leading side of the
earth's orbit.

Full Moon
The moon is opposite
from the sun, on the far
side of the earth.

Waxing Gibbous Moon
The moon is moving into
the far side of the earth's
orbit, away from the sun.

First Quarter Moon
The moon is 90 degrees
(1/4) around the earth,
still on the trailing side.

Young Crescent Moon
The moon is trailing
behind the earth in its
orbit.

New Moon
The moon is between the sun and the earth.

the first quarter moon when the terminator continues to move to the right, lighting more and more of the surface, is still part of the waxing moon. It is also sometimes referred to as the gibbous moon (which can also refer to the period after the full moon up until the last quarter moon), or more specifically, the waxing gibbous moon.

The moon becomes full when the orbital path of the moon carries it directly opposite of the sun, on the other side of the earth. At this position, it receives the direct light of the sun across the full face, forming the distinctive full moon. Another term for full moon is "moon in opposition," because it is opposite the sun. Being opposite the sun at the time of full moon, the moon rises just as the sun is setting.

The full moon immediately begins diminishing, even though it might not be visibly apparent for a day or two, because the spherical shape of the moon can hide the first effects of the growing shadow on the right-hand side. The period after the full moon up until the new moon is called a waning moon. The period between the full moon and the last quarter moon is more specifically called a waning gibbous moon. During this part of the cycle, the sun is catching up to the moon. The sunlight that illuminates it is now coming from "behind," so the lighted surface starts to diminish from the right.

The growing shadow on the right-hand side of the face gradually increases until it forms an almost perfect half circle. At this point, the phase is called the last quarter moon. It is opposite to the first quarter moon, with the lighted half now being on the left side. Like the first quarter moon, this phase is also technically referred to as a quadrature, and the moon is now positioned at a right angle (perpendicular) to the imaginary line connecting the sun and the earth.

The moon continues to "shrink" after the last quarter, with the growing shadow obsuring more and more of the face. In less than a week, the entire face will be in shadow, and the cycle will be back at the beginning, the new moon. The last part of the visible phase is often called the old crescent moon.

Lunar Phases

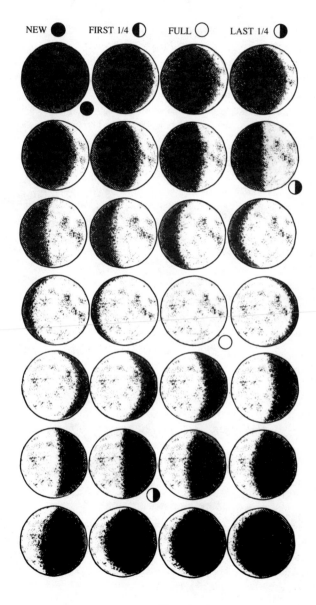

NEW   FIRST 1/4   FULL   LAST 1/4

31

Comparing Moon Phases

Days in Lunation	Common Name of Phase	Time of Day
0 days	New Moon *(rises and sets with the sun)*	
1	Waxing Crescent *(lags a few hours behind the sun)*	
2		
3		
4		
5		
6		
7.4	First Quarter *(above horizon 1/2 in day, 1/2 at night)*	
8	Waxing Gibbous *(lags 8-10 hours behind the sun)*	
9		
10		
11		
12		
13		
14.8	Full Moon *(rises about sunset, sets about sunrise)*	
15	Waning Gibbous *(precedes sun by 8-10 hours)*	
16		
17		
18		
19		
20		
21		
22.1	Last Quarter *(above horizon 1/2 at night, 1/2 day)*	
23	Waning Crescent *(precedes sun by a few hours)*	
24		
25		
26		
27		
28		
29.5	New Moon	

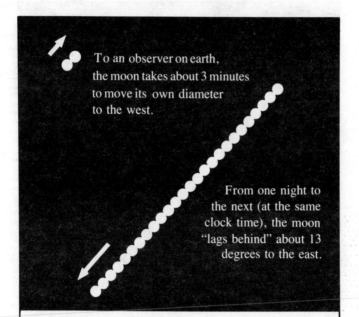

To an observer on earth, the moon takes about 3 minutes to move its own diameter to the west.

From one night to the next (at the same clock time), the moon "lags behind" about 13 degrees to the east.

Speed of the Moon

The apparent motion of the moon across the sky is mostly the result of the rotation of the earth. The speed of the earth's rotation accounts for about 95 percent of the moon's visible motion. The other 5 percent is from the moon's actual movement.

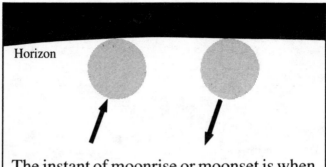

The instant of moonrise or moonset is when the upper limb of the moon is even with the horizon.

Moonrise and Moonset

The moon rises and sets at different times every day because the calendar is based on a solar timetable, not a lunar one. On the average, moonrise and moonset are about one hour later each succeeding day, but the times change considerably from one location to another. Both latitude and longitude have an effect on this change. In some northern latitudes (northern Canada, for instance), there is a dramatic change from day to day in the times for moonrise and moonset. During some periods of the lunar cycle, there are many days when the moon does not rise at all, or the opposite, does not set.

During each lunar month, there is one day with no moonrise and one day with no moonset. This happens because of the lagging behind of the moon compared to the 24-hour day. The moon actually has a 25-hour day (approximately). Therefore, for example, if the moon set at 11:50 P.M. on a Tuesday night, 25 hours later would be completely past Wednesday, and the next set time would be about 12:40 A.M. on Thursday morning.

MOONRISE AND MOONSET, 1989
UNIVERSAL TIME FOR MERIDIAN OF GREENWICH
MOONRISE

Lat.		+40°	+42°	+44°	+46°	+48°	+50°	+52°	+54°	+56°	+58°	+60°	+62°	+64°	+66°
		h m	h m	h m	h m	h m	h m	h m	h m	h m	h m	h m	h m	h m	h m
Oct.	24	1 19	1 16	1 13	1 09	1 06	1 02	0 57	0 53	0 47	0 41	0 35	0 27	0 18	0 07
	25	2 20	2 18	2 17	2 15	2 14	2 12	2 10	2 07	2 05	2 02	1 59	1 56	1 51	1 47
	26	3 20	3 20	3 20	3 20	3 20	3 20	3 20	3 20	3 20	3 21	3 21	3 21	3 21	3 22
	27	4 19	4 20	4 22	4 24	4 26	4 28	4 30	4 32	4 35	4 38	4 42	4 46	4 50	4 56
	28	5 18	5 21	5 25	5 28	5 32	5 36	5 40	5 45	5 51	5 57	6 04	6 12	6 21	6 32
	29	6 19	6 23	6 28	6 33	6 39	6 45	6 52	7 00	7 08	7 18	7 29	7 42	7 57	8 16
	30	7 21	7 26	7 33	7 40	7 47	7 56	8 05	8 15	8 27	8 41	8 57	9 16	9 41	10 16
	31	8 23	8 30	8 37	8 46	8 55	9 06	9 17	9 30	9 46	10 04	10 27	10 56	11 42	-- --
Nov.	1	9 23	9 31	9 40	9 50	10 00	10 12	10 26	10 42	11 00	11 23	11 53	12 39	-- --	-- --
	2	10 20	10 29	10 38	10 48	10 59	11 12	11 26	11 43	12 03	12 28	13 02	14 05	-- --	-- --
	3	11 11	11 19	11 28	11 38	11 49	12 01	12 15	12 31	12 50	13 13	13 43	14 30	-- --	-- --
	4	11 55	12 02	12 10	12 19	12 29	12 39	12 51	13 05	13 21	13 39	14 02	14 32	15 18	-- --
	5	12 32	12 38	12 45	12 52	13 00	13 09	13 18	13 29	13 41	13 55	14 11	14 31	14 55	15 29
	6	13 04	13 09	13 14	13 19	13 25	13 31	13 38	13 46	13 55	14 04	14 15	14 28	14 43	15 02
	7	13 32	13 35	13 38	13 42	13 46	13 50	13 54	13 59	14 05	14 11	14 18	14 25	14 34	14 45
	8	13 58	13 59	14 01	14 03	14 04	14 06	14 08	14 11	14 13	14 16	14 19	14 23	14 27	14 31
	9	14 23	14 23	14 23	14 22	14 22	14 22	14 22	14 21	14 21	14 21	14 20	14 20	14 19	14 18
	10	14 50	14 48	14 46	14 43	14 41	14 38	14 36	14 33	14 29	14 26	14 22	14 17	14 12	14 06
	11	15 19	15 15	15 11	15 07	15 03	14 58	14 52	14 46	14 40	14 32	14 24	14 14	14 03	13 51
	12	15 54	15 48	15 43	15 36	15 29	15 22	15 13	15 04	14 54	14 42	14 29	14 13	13 54	13 29
	13	16 37	16 30	16 22	16 14	16 05	15 54	15 43	15 31	15 16	14 59	14 39	14 13	13 38	12 34
	14	17 30	17 21	17 12	17 03	16 52	16 40	16 26	16 11	15 53	15 30	15 02	14 20	** **	** **
	15	18 32	18 24	18 15	18 05	17 53	17 41	17 27	17 10	16 51	16 27	15 54	15 01	** **	** **
	16	19 42	19 34	19 26	19 16	19 06	18 55	18 42	18 27	18 10	17 49	17 23	16 45	** **	** **
	17	20 53	20 47	20 40	20 32	20 24	20 14	20 04	19 53	19 39	19 24	19 05	18 42	18 10	17 19

MOONSET

Lat.		+40°	+42°	+44°	+46°	+48°	+50°	+52°	+54°	+56°	+58°	+60°	+62°	+64°	+66°
		h m	h m	h m	h m	h m	h m	h m	h m	h m	h m	h m	h m	h m	h m
Oct.	24	14 46	14 48	14 50	14 52	14 55	14 58	15 00	15 04	15 07	15 11	15 15	15 20	15 26	15 32
	25	15 07	15 07	15 08	15 09	15 09	15 10	15 11	15 12	15 13	15 14	15 16	15 17	15 19	15 21
	26	15 27	15 26	15 25	15 24	15 23	15 22	15 21	15 20	15 19	15 17	15 16	15 14	15 12	15 09
	27	15 47	15 45	15 43	15 40	15 38	15 35	15 32	15 28	15 25	15 20	15 16	15 10	15 04	14 58
	28	16 09	16 06	16 02	15 58	15 53	15 49	15 44	15 38	15 31	15 24	15 16	15 07	14 57	14 44
	29	16 34	16 29	16 24	16 18	16 12	16 05	15 58	15 50	15 40	15 30	15 18	15 04	14 48	14 28
	30	17 03	16 57	16 50	16 43	16 35	16 26	16 16	16 05	15 53	15 39	15 22	15 02	14 36	14 00
	31	17 38	17 30	17 22	17 13	17 04	16 53	16 41	16 28	16 12	15 53	15 30	15 00	14 14	-- --
Nov.	1	18 20	18 11	18 02	17 53	17 42	17 30	17 16	17 00	16 41	16 18	15 48	15 02	-- --	-- --
	2	19 10	19 01	18 52	18 42	18 31	18 18	18 03	17 47	17 27	17 02	16 28	15 25	-- --	-- --
	3	20 08	20 00	19 51	19 41	19 30	19 18	19 05	18 49	18 30	18 08	17 38	16 51	-- --	-- --
	4	21 12	21 05	20 57	20 48	20 39	20 29	20 17	20 04	19 49	19 30	19 08	18 39	17 54	-- --
	5	22 19	22 14	22 07	22 01	21 53	21 45	21 36	21 26	21 15	21 01	20 46	20 27	20 03	19 31
	6	23 29	23 25	23 20	23 16	23 10	23 05	22 58	22 51	22 44	22 35	22 25	22 13	21 59	21 42
	7	.. ..	.. ..	.. ..	.. ..	.. ..	.. ..	.. ..	.. ..	.. ..	.. ..	.. ..	23 56	23 49	23 40
	8	0 40	0 37	0 35	0 32	0 29	0 25	0 22	0 18	0 13	0 08	0 03	.. ..	.. ..	.. ..
	9	1 51	1 51	1 50	1 49	1 48	1 48	1 47	1 45	1 44	1 43	1 41	1 40	1 38	1 35
	10	3 05	3 07	3 08	3 09	3 10	3 12	3 14	3 16	3 18	3 20	3 22	3 25	3 29	3 32
	11	4 22	4 25	4 29	4 32	4 36	4 40	4 44	4 49	4 55	5 01	5 08	5 16	5 26	5 37
	12	5 42	5 47	5 53	5 58	6 05	6 11	6 19	6 27	6 37	6 48	7 00	7 15	7 33	7 56
	13	7 04	7 11	7 18	7 26	7 34	7 44	7 55	8 07	8 21	8 37	8 57	9 22	9 56	11 00
	14	8 23	8 31	8 40	8 49	9 00	9 12	9 25	9 40	9 58	10 20	10 49	11 30	** **	** **
	15	9 34	9 42	9 52	10 02	10 13	10 26	10 40	10 56	11 16	11 40	12 12	13 06	** **	** **
	16	10 33	10 41	10 50	10 59	11 09	11 21	11 34	11 49	12 07	12 28	12 55	13 35	** **	** **
	17	11 19	11 26	11 34	11 42	11 50	12 00	12 11	12 23	12 37	12 53	13 12	13 37	14 09	15 00

(.. ..) indicates phenomenon will occur the next day.
(-- --) indicates Moon continuously below horizon.
(** **) indicates Moon continuously above horizon.

Sample page from the Astronomical Phenomena *(see page 114)*, which lists times for the lunar cycles.

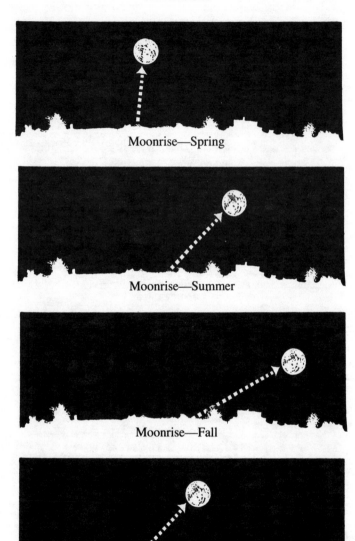

Moonrise—Spring

Moonrise—Summer

Moonrise—Fall

Moonrise—Winter

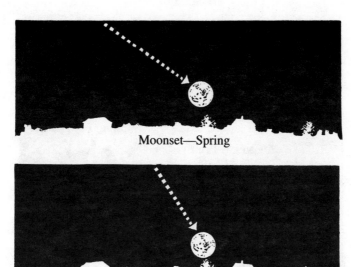

Moonset—Spring

Moonset—Summer

Moonset—Fall

Moonset—Winter

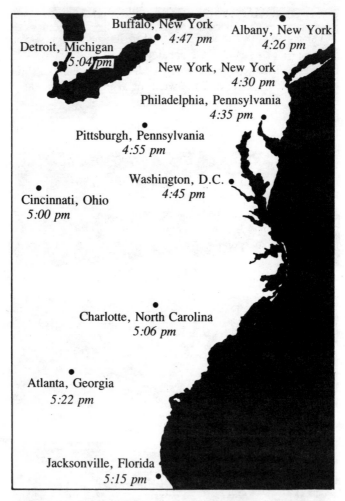

The time of moonrise is affected by latitude and longitude. The map (eastern U.S.) shows various locations and the times of moonrise for October 24, 1988.

Moon Size and Moonlight

The apparent size of the moon to observers on earth is about 1/2 degree. This is also the apparent size of the sun. If you hold your hand up to the moon with arm outstetched, this corresponds to about half the width of a finger. Sometimes, however, the moon appears to be much larger. When rising and setting, for example, the moon at full phase often appears to be larger than when it is directly overhead.

There are factors which can change the diameter, and brightness, of the moon, but the rising and setting phenomena are actually optical illusions. At these times, the closeness to the horizon adds a remarkable degree of visual influence. The eye is tricked into measuring the moon against nearby objects, and the visual elements found on most horizons—buildings, trees, hills— create the impression of increased size.

This optical trick can be demonstrated by using cardboard masks to cut off the view of nearby objects. The viewing hole should be no larger than 1/4 inch in diameter. Tape the mask to a yardstick so that it can be kept at a fixed distance from your eye. As a full moon rises, view it through the mask when it just clears the horizon. Note the size of the moon's disk compared to the size of the hole, then wait for the moon to reach its zenith and view it again.

The view of the moon from the horizon (A) is further—by ½ the diameter of the earth —than when the moon is overhead (B).

   Although most of the apparent size change in the moon is caused by visual tricks, there are physical effects which cause a measurable size difference, although not a large one. One of these effects is the difference in distance between an observer and the moon from a horizontal perspective to a vertical one. This difference—the change from viewing the moon on the horizon and at its zenith—is half the diameter of the earth. However, this accounts for only a maximum of 2 percent difference in the size of the moon, hardly enough to notice with the naked eye. And in any case, this size difference decreases the size of the moon at the horizon, not the opposite, because that is when it is farthest away from the observer.

A more significant difference is caused by the variation in the moon's orbit around the earth. This orbit is not round, but elliptical (oval) in shape, making some points in the orbit closer to the earth than others. At the closest point (perigee), the moon will appear measurably larger than when at the farthest point (apogee). The difference in diameter from one to the other is about 14 percent.

Moon at apogee   Moon at perigee
(farthest from earth)   (closest to earth)

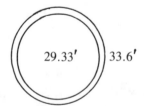

14% difference in size

The moon makes one orbit every 29.5 days, so there is one perigee and one apogee during every orbital period. However, the phases of the moon are not synchronized with this cycle, and a full moon—the most visible phase—will occur on or near a perigee or apogee in only a few months in any calendar year. If the two cycles coincided in August, with the full moon occurring on about the same day (give or take few days) as the perigee, they would also coincide in September, October, and November, with the date of perigee gradually "drifting" backwards through the calendar month faster than the date of full moon. By December or January, the date of perigee would be more than a few days before the full moon, and the distance effect would thus be diminished.

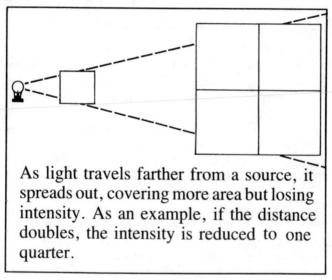

As light travels farther from a source, it spreads out, covering more area but losing intensity. As an example, if the distance doubles, the intensity is reduced to one quarter.

The same kind of distance factors which change the size of the moon can affect the amount of sunlight it reflects. Light is affected by the law of inverse squares—it decreases as the square of the distance. This happens because light spreads out as it gains distance; the farther from the source, the more area that the same amount of light is covering. For example, if you double the distance between you and a light bulb from 10 to 20 feet, the illumination would drop by 75 percent.

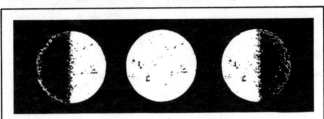

The first and last quarter moons are only 1/9th as bright as the full moon.

The moon does not alter its distance from the sun by much, but there is enough change to affect the intensity of light it receives. Also, the changing distance between the moon and earth affects the amount of sunlight reflected from the moon to earth. However, these changing values are rarely noticeable because of the interfering effects of the earth's atmosphere, which diffuses the moonlight.

The greatest amount of moonlight comes during a full moon. At the highest point in the sky, a full moon gives about 0.02 foot-candles illumination. Quarter moons provide only about 1/9th of the light of a full moon. The first quarter moon is about 20 percent brighter than the last quarter moon. These differences come from the spherical shape of the moon and the coarse, uneven surface. Sunlight striking the full face of the moon, as seen during a full moon, produces a maximum reflection because it is being reflected directly back toward the earth. At any other time during the phase cycle, sunlight is striking the surface obliquely, limiting its ability to be reflected directly back at the earth.

Rules of Thumb

The apparent size and brightness of the moon is determined by a combination of factors, including:

■ The closeness of the moon to the horizon. Especially during full moons, the moon appears much larger when it is just rising or setting because the proximity of the horizon acts as a visual guide to its size. When the moon is high in the sky, there is nothing but stars and planets for comparison, both of which are just points of light next to it.

■ The angle of the moon's path compared to the horizon. This angle is greatest after sunset in the spring months and before sunrise in fall months. The angle is smallest before sunrise in the spring months and after sunset in fall months. When the angle is smallest, the moon stays closer to the horizon for a longer period of time, adding to the impression of increased size.

■ The distance of the moon from the earth. At greatest distance (around the apogee), the moon appears smaller, at closest distance (around the perigee), it appears larger.

■ The speed of the moon in its orbit around earth. At greatest speed (around the time of perigee) the phases appear to change more quickly; at slowest speed (around the time of apogee), the phases appear to change more slowly.

■ The variation in brightness of full moons (not including effects of atmosphere or light pollution) is about 10 percent.

■ The distance of the moon from the sun. At greatest distance (around the time of aphelion of earth orbit) the moon is dimmer than when it is closest (around the time of perihelion of earth orbit).

Lunar Eclipses

Lunar eclipses are caused when the earth's shadow falls on the moon. This would happen every full moon if the moon orbited around the earth in the same plane as the earth orbits around the sun. The moon's orbit, however, is tilted about 5 degrees above the earth-sun plane. Lunar eclipses happen only when the tilt of this plane lines up with the earth-sun plane.

A lunar eclipse is visible at the same time to everyone who is in sight of the full moon. This type of eclipse can last for more than three hours because the moon and the earth are moving slowly in relation to each other, and the shadow cast by the earth is so large. This shadow is much larger than that cast by the moon on the earth (during a solar eclipse) because of their sizes and the relative distances between the earth, moon, and sun.

There are three kinds of eclipses of the moon.

1. Penumbral eclipse. This is a partial eclipse, with the moon in only the secondary shadow (penumbra) of the earth. A penumbral eclipse is sometimes called an appulse eclipse. During a penumbral eclipse, the moon's light is dimmed, but it does not go dark, because the penumbral shadow is not dark enough to block much of the sun's illumination. Often, there

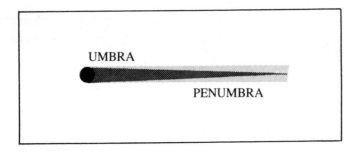

is no visible line separating the shadow from the sunlight on the moon's surface, and the eclipse is only noticeable as a slight darkening of the lunar surface.

Lunar Eclipse Sequences

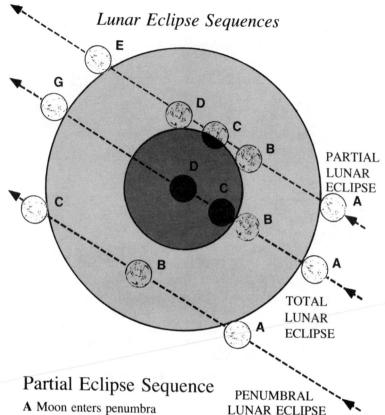

PARTIAL
LUNAR
ECLIPSE

TOTAL
LUNAR
ECLIPSE

PENUMBRAL
LUNAR ECLIPSE

Partial Eclipse Sequence

A Moon enters penumbra
B Moon enters umbra
C Middle of eclipse
D Moon leaves umbra
E Moon leaves penumbra

Penumbral Eclipse Sequence

A Moon enters penumbra
B Middle of eclipse
C Moon leaves penumbra

Total Eclipse Sequence

A Moon enters penumbra
B Moon enters umbra
C Total eclipse begins
D Middle of eclipse
E Total eclipse ends
F Moon leaves umbra
G Moon leaves penumbra

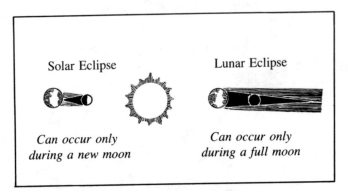

Solar Eclipse

Lunar Eclipse

*Can occur only
during a new moon*

*Can occur only
during a full moon*

2. Partial eclipse. The moon partially enters the main shadow (umbra) of the earth.

3. Total eclipse. The moon is completely inside the main shadow (umbra) of the earth. The dark umbral shadow cast by the earth usually does not completely obscure the moon but changes its color to a dull copper tone, an effect created by earthshine (light reflected off the earth onto the moon). The color is created by the filtering effect of the earth's atmosphere, which removes all but the red wavelengths of sunlight. The moon can stay in the umbral shadow of the earth for as long as 90 minutes. The movement through the penumbral shadow can last for about 60 minutes.

Solar Eclipses

The moon also causes eclipses of the sun. When the new moon comes directly between the earth and sun, it blocks out the sun's rays. Depending on how the moon and the sun line up, there can be either partial or total blocking of the sun's disk.

A total solar eclipse occurs when the moon completely blocks out the sun. However, the elliptical orbit of the moon can place it at varying distances from the earth. When the orbit is closer to the earth during a solar eclipse, the moon appears larger

Solar Eclipses

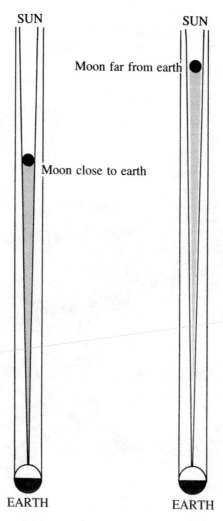

Total eclipse

Annular eclipse

47

and therefore blocks out the sun for a longer period of time. When the orbit is farther from the earth during an eclipse, the moon appears smaller and does not quite cover the sun's disk. This event is referred to as an annular eclipse.

A solar eclipse is much shorter in duration than a lunar eclipse because the moon's shadow is falling on a rapidly rotating earth. The maximum time for a solar eclipse is 7 minutes and 40 seconds (on the equator) but most of these eclipses are much shorter. The moving shadow cast by the eclipse is called an eclipse track and is usually about 3,000 miles long and 100 miles wide. The width can vary from almost nothing to a maximum of 167 miles wide.

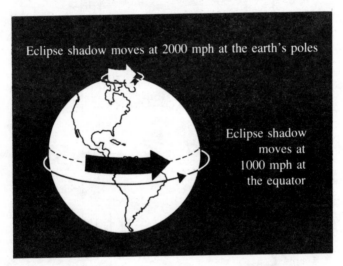

The moon's shadow during a solar eclipse moves along at speeds between 1,000 and 2,000 miles per hour. The slowest movement is at the equator and the fastest at the poles. The speed of the shadow is caused by the combined movements of the moon and the earth, the former moving at about 2,000 miles per hour eastward, and the former (at the equator) rotating at about 1,000 miles per hour.

There are three kinds of eclipses of the sun.

1. Partial eclipse. The moon is in conjunction with the sun but its path does not take it directly across the center of the sun's disc.

2. Annular eclipse. The moon is in conjunction with the sun, but is at a far point in its orbit around the earth. The moon's image is thus too small to completely cover the sun's disc, and a ring of sunlight is visible around the edges.

3. Total eclipse. The moon is in conjunction with the sun and is at a close point in its orbit around the earth. The moon's image is thus large enough to completely block out the sun's disc.

Eclipse Rules of Thumb

■ Full moons are the only time when lunar eclipses can occur.

■ New moons are the only time when solar eclipses can occur.

■ Lunar eclipses last for as long as 3 hours and 40 minutes, with the period of totality lasting for as long as 1 hour and 40 minutes.

■ Solar eclipses can last for as long as 7 minutes and 40 seconds (when visible at the equator) if they are total, 12 minutes and 24 seconds at most if they are annular.

■ Solar eclipses happen at least twice a year but never more than five times a year.

■ There can never be more than three lunar eclipses a year.

■ The most number of solar and lunar eclipses that can happen in a year is seven.

■ Solar eclipses and lunar eclipses go together in pairs. A solar eclipse is always followed or preceded by a lunar eclipse, within an interval of 14 days.

■ A sequence of eclipses is repeated every 18 years, 11 days, and 8 hours in what is called the Saros cycle.

Earthshine

The light of the sun reflects not only off the face of the moon, it also reflects off the earth's surface. The reflected sunlight can produce visible light on the moon. During periods when the moon is lit directly by sunlight, earthshine is usually not visible. During eclipses and occasionally just after a new moon, however, the earthshine effect can be seen. It is visible as a dull copper or reddish hue. Total lunar eclipses are the only eclipses exhibiting this phenomenon.

Illumination of the moon just after the new moon produces an effect sometimes referred to as the "old moon in the new moon's arms." This phenomenon is visible just after sunset a few days after the new moon when the young crescent moon is beginning to set in the west. The thin crescent-shaped illumination on the right side of the moon's face will be seen, along with the rest of the surface dimly visible with a copper or reddish hue. For early risers, the same effect is sometimes visible as the old crescent moon sets just before sunrise.

The Moon is Flat?

Early observers of the moon, including Galileo, noticed that the full moon appeared to be flat. If a round object such as a balloon or ball is illuminated like the full moon, a distinct three-dimensional effect occurs. In fact, almost any sphere exhibits this common effect from illumination—a gradual darkening around the edges with the brightest area in the center. The moon does not.

That the full moon has almost equal illumination everywhere across its surface is the result of its unusual surface texture. Light from the sun striking the moon is almost completely absorbed by the surface. Only about 12 percent of the sunlight is reflected. The percent of light reflected from an object is referred to as albedo; therefore the moon's albedo is 12 (0.12). In comparison, the albedo of earth is about 37 (37 percent, or

0.37); for Venus it is 65 (0.65); and for Mars it is 15 (0.15).*

The texture and color of different substances affects albedo. For instance, on earth the albedo of concrete ranges from 17 to 27; the albedo of snow is 45 to 90; the albedo of deserts is 25 to 30 percent; and the albedo of soil is 5 to 15 percent. Various features on the moon also have different albedos. This results from the primary materials present at each location. The range is from 0.05 (Sinus Medii) to 0.176 (Aristarchus). Even during eclipses, the lunar features with the highest albedos can be seen because of earthshine.

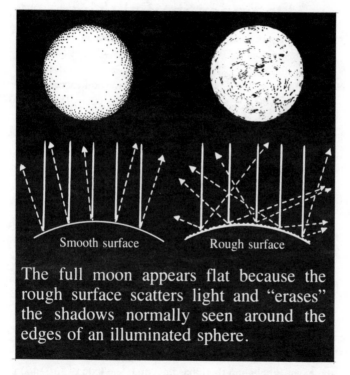

Smooth surface   Rough surface

The full moon appears flat because the rough surface scatters light and "erases" the shadows normally seen around the edges of an illuminated sphere.

* Figures from the *Observer's Handbook, 1988* (Royal Astronomical Society of Canada).

The almost complete absorption of light on the moon is caused by the rough, uneven texture of the surface at measurements below a centimeter (less than 1/2 inch). The surface appears smooth, but the "micro-roughness" of the surface, which was not measurable until spacecraft landed, does an efficient job of scattering sunlight into an infinite number of directions. Around the edges of the illuminated disk of the full moon, this redirection of light eliminates the shadows which would ordinarily be seen on any spherical object.

Observing the Full Moon

The full moon occurs at a precise moment in time. In astronomical terms, it is full at the moment when it is directly opposite the sun. If that moment were exactly at the time of sunset for your location, you would see the full moon rise in the east just as the sun was setting in the west. This rarely happens, however. What usually occurs is a slight discrepancy in the time at which the moon is full and the time it rises at your location. The consequence is that the fully lit moon will rise from a few minutes to an hour before or after sunset. The moon, however, will appear full for at least a day—and sometimes for more than a day—because its spherical shape does not begin to show the waning shadow (on the right side) for one to two days after the full moon.

On the marked day of a full moon, no matter what the astronomical time of the event, the moon will appear full during the evening hours closest to the astronomical time of the phase. In general, if the marked time is 9:00 A.M. or later, you will see the fullest moon on that evening. If the marked time is earlier, you will see the fullest moon on the previous evening.

No matter where you are on the globe, the time listed as Greenwich Mean Time (GMT) for any phase of the moon is the exact time when that phase occurs. The time of full moon and new moon is usually listed in calendars and almanacs in

Greenwich Mean Time. This is the official local time at the Greenwich Meridian (near London, England) which is used by astronomers for consistency. There are no variations in latitude or longitude that affect the time that phases occur, but if you are not in the GMT zone, you do have to convert that time to your local time zone. In order to convert GMT to your local time, you must add or subtract the right number of hours, determined by how many time zones are between you and Greenwich, England. *(See chart on page 109.)*

Lunar Tilt

The moon appears to be tilting, or changing position, as it crosses the sky. This effect is most visible when the moon is waxing, and the lighted portion of the surface appears to turn from pointing to the west (right) to almost straight down as the moon sets. The moon is not actually turning, however, but is following a curved path across the sky, the path defining its orbit. Pictures or illustrations of the lunar phases usually show the moon when it is at its zenith (highest point above an observer) and therefore oriented straight "up and down" relative to the earth.

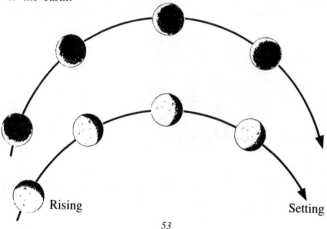

Rising   Setting

Hide and Seek

The sky is full of visible objects to observers on earth. Planets and stars are visible as points of light to the naked eye. The planets appear to move at different rates than the stars (the apparent movement of the stars is actually due to the movement of the earth) as they are much closer to the earth. A star or a planet will occasionally be obscured by the moon when their paths cross in the sky. This phenomenon is called an occultation.

Occultations occur whenever one body crosses in front of another body in the sky. Occultations caused by the moon are the most obvious, as the moon's image is so large. The most interesting occultations to watch are those that occur when the moon is less than full, as the other objects are then less likely to be lost in the light of the moon.

Observers on earth see an occultation as a point of light approaching the moon from the right. The stars and planets move across the sky faster than the moon, accounting for this approach. Occultations can last for a few minutes or up to an hour, depending on the moon's path (declination), that is, whether the object appears to intersect the moon at the widest part or just "grazes" the edge. At the instant of occultation,

Grazing occultation on dark limb

Grazing occultation on bright limb

Two types of grazing occultations.

the point of light will suddenly disappear. It will then "blink" on again on the other side of the moon. Grazing occultations appear as a flashing or twinkling phenomena, with the mountains and ridges on the moon's surface momentarily breaking up the light from the star.

During a grazing occultation, the moving point of light from a star will "blink" on and off as it moves behind the peaks and valleys on the lunar surface.

The brighest and therefore most visible stars that are likely to be observed being occulted by the moon are Aldebaran, Antares, Regulus, the Pleiades, and Spica. These stars are all located in the band of sky across which the moon travels, a 5 degree swath on either side of the ecliptic.

Lunar occultations are predictable because of the known orbital characteristics of the moon. However, the times of occulations are difficult to compute because of the widely varying path of the lunar orbit and the range of possible positions of observers. Amateur astronomers who are serious about observing lunar occultations can receive specific information for their locations from the U.S. Naval Observatory. In return they are expected to keep notes on their observations and share this information with the Observatory. Serious inquiries about this program can be addressed to:

U.S. Naval Observatory
Washington, D.C. 20390

Librations

The moon always shows the same face to the earth. Earth observers, however, get to see more than half of the surface of the moon because of phenomena called librations. Librations allow the front surface of the moon to be seen from slightly different angles at different times, producing an overall picture of the lunar surface that adds up to 59 percent of the total. At any one time, the most that can be seen of the surface is only 41 percent, because the spherical shape of the moon hides the area close to the perimeter. The visible effect of librations is usually not noticeable to the naked eye, but creates a fascinating panorama of changing vistas through binoculars or telescopes when observed over a period of lunations.

Librations are measured using longitudinal and latitudinal coordinates. Both are figured from a central point which is at a fixed geographical location on the lunar surface. This point is in Sinus Medii, a small flat plain just below and to the right of Copernicus, a large rayed crater that is visible to the naked eye.

Libration in Longitude. This libration is produced by the elliptical orbit of the moon. If the orbit were a circle, the moon's face would always point directly at the earth and its orbital speed would remain the same. However, due to the nature of an elliptical orbit, the speed of the moon changes according to which part of the orbit it is in. This is also true of any object moving in an elliptical orbit. When moving from its fastest point (closest to earth) to its slowest point (farthest from earth), the moon's speed is slowing down, but its rotation speed remains the same. For a period of time, the face of the moon is therefore not pointed directly at the earth. This "lag" effect allows observers to see an extra slice of surface, in effect, to "peek" around the edge of the moon. When the moon is at the 90 degree point in its revolution (one-quarter of the way around the earth), it is 97 degrees through its rotation. This libration is called longitudinal because the extra surface areas

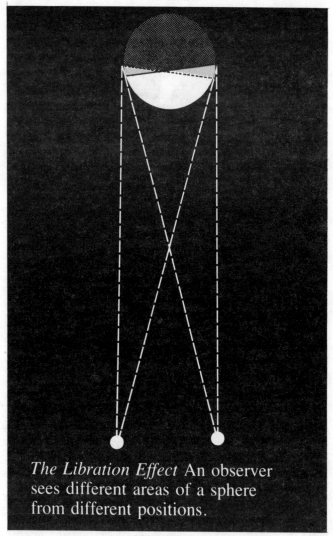

The Libration Effect An observer
sees different areas of a sphere
from different positions.

are exposed along lines of longitude (perpendicular to the
equator). The total extra surface that can be seen with this
libration is 8 degrees.

Libration in Longitude

Libration in Latitude. The plane of the moon's orbit is tilted about 5 degrees away from the plane of the earth's orbit around the Sun. For half of a lunar cycle, the moon is below the ecliptic and for half the cycle it is above the ecliptic. Each of these half cycles exposes an extra "slice" of the lunar surface at the top of its northern hemisphere or the bottom of its southern hemisphere, in effect allowing a "peek" above or below the normal limit of visible surface. This libration is called latitudinal because the extra surface areas that are exposed are great circles that are parallel to the equator. The total extra surface that can be seen with this libration is about 7 degrees.

Libration in Latitude

Diurnal Libration. Observers can "see over the top" of the moon when it is rising, and "under the bottom" when it is setting. This is possible because the radius of the earth adds an extra 4,000 miles of height advantage for looking "over" or "under" the moon when it is on the horizon. This libration is called diurnal because it occurs every day, but only accounts for an extra 1 degree of visible surface.

Diurnal Libration

Tides

The moon is the nearest celestial neighbor to earth and exerts its influence through gravitational attraction. Although this force is only one ten-millionth of the gravitational force of the earth, it is enough to create tides on the world's bodies of waters. There are also tides created in the atmosphere and to a much lesser degree in the earth itself.

Lunar gravity does not work alone in producing tides, they are also influenced by the centrifugal force created from the earth's revolution around its barycenter *(see page 122)* and the gravitational attraction of the sun. Gravitational attraction varies inversely with the square of the distance between two bodies. The sun, for example, is 27 million times larger than the moon, but it is 390 times farther away from the earth than the moon. The result is that the sun's gravitational force on the earth is only 46 percent as much as the moon's, making the moon the most important factor in creating tides.

Tides are not created by the direct pull of the moon's gravity, which is not strong enough in itself to make a measurable tide. The strongest forces in producing tides are called tractive forces, horizontal forces created by the moon's pull on water that is not directly underneath it. These tractive forces create a "piling" effect which adds up to a bulging mass of water somewhat behind the actual zenith of the moon. The time it takes for this high water to arrive after the moon passes overhead is called the lunitidal interval. At the same time that water has piled up on the moon's side of the earth, a second bulge has piled up on the opposite side of the planet. This second bulge is a result of water moving to create an equilibrium from the forces created by the movement of the earth and the pull of the moon and sun. As the two bulges move around the earth, the earth itself is moving under them, further complicating the motion of the tides.

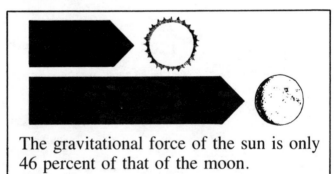

The gravitational force of the sun is only 46 percent of that of the moon.

Local geography also complicates the motion of tides. The actual speed and height of the tide is slowed by such things as the depth of the water, obstructions to the tides' progress such as land masses, and wind. In the open ocean, with deep water and no land obstructions, the difference between high and low tides is about one foot, but in some coastal locations, the difference can be more than twenty feet.

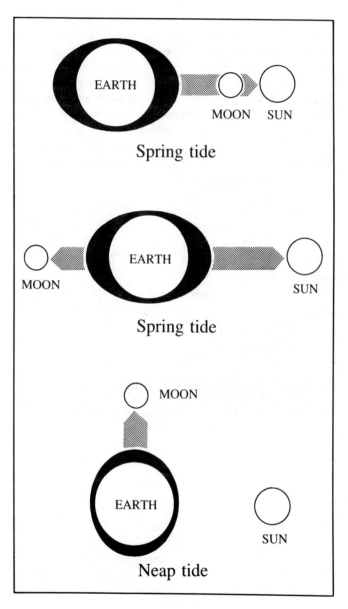

Spring tide

Spring tide

Neap tide

When sun, moon, and earth are in a line, the maximum monthly tides occur. If the sun and moon are on the same side of the earth (new moon) or on opposite sides (full moon), a spring tide is the result. Spring tides create the highest high waters and lowest low waters every month. When the sun and moon are at right angles (first and last quarter moons), a neap tide is the result, and the difference between high and low waters is at a minimum.

Sample tide tables from the National Oceanic and Atmospheric Administration, National Ocean Survey.

SAN FRANCISCO (Golden Gate), CALIFORNIA, 1981

Times and Heights of High and Low Waters

APRIL

Day	Time h.m.	Height ft.	Height m.
1 W	0227	2.0	0.6
	0827	5.3	1.6
	1449	-0.4	-0.1
	2153	4.9	1.5
2 Th	0311	1.4	0.4
	0923	5.4	1.6
	1532	-0.4	-0.1
	2227	5.2	1.6
3 F	0355	0.7	0.2
	1015	5.5	1.7
	1614	-0.3	-0.1
	2259	5.4	1.6
4 Sa	0436	0.1	0.0
	1109	5.5	1.7
	1656	0.0	0.0
	2334	5.7	1.7
5 Su	0525	-0.4	-0.1
	1204	5.3	1.6
	1735	0.4	0.1
6 M	0010	5.9	1.8
	0611	-0.7	-0.2
	1300	5.1	1.6
	1820	1.0	0.3
7 Tu	0051	6.0	1.8
	0703	-0.8	-0.2
	1402	4.8	1.5
	1907	1.5	0.5
8 W	0134	6.0	1.8
	0758	-0.8	-0.2
	1509	4.6	1.4
	2001	2.1	0.6
9 Th	0224	5.9	1.8
	0901	-0.6	-0.2
	1624	4.5	1.4
	2107	2.5	0.8
10 F	0322	5.6	1.7
	1007	-0.4	-0.1
	1744	4.5	1.4
	2229	2.7	0.8
11 Sa	0424	5.4	1.6
	1118	-0.2	-0.1
	1853	4.6	1.4
	2353	2.6	0.8
12 Su	0541	5.1	1.6
	1226	-0.1	0.0
	1954	4.8	1.5
13 M	0106	2.3	0.7
	0653	5.0	1.5
	1325	0.0	0.0
	2041	5.0	1.5
14 Tu	0204	1.8	0.5
	0759	4.9	1.5
	1418	0.1	0.0
	2120	5.1	1.6
15 W	0254	1.4	0.4
	0858	4.9	1.5
	1500	0.2	0.1
	2155	5.2	1.6
16 Th	0336	0.9	0.3
	0950	4.8	1.5
	1539	0.4	0.1
	2226	5.2	1.6
17 F	0412	0.6	0.2
	1037	4.7	1.4
	1614	0.7	0.2
	2254	5.2	1.6
18 Sa	0447	0.3	0.1
	1121	4.5	1.4
	1646	1.0	0.3
	2318	5.1	1.6
19 Su	0519	0.1	0.0
	1204	4.4	1.3
	1717	1.4	0.4
	2344	5.1	1.6
20 M	0554	0.0	0.0
	1246	4.3	1.3
	1749	1.8	0.5
21 Tu	0007	5.1	1.6
	0630	-0.1	0.0
	1332	4.2	1.3
	1821	2.2	0.7
22 W	0034	5.1	1.6
	0705	-0.1	0.0
	1418	4.1	1.2
	1857	2.6	0.8
23 Th	0105	5.1	1.6
	0744	0.0	0.0
	1514	4.0	1.2
	1939	2.9	0.9
24 F	0144	5.0	1.5
	0829	0.1	0.0
	1613	4.0	1.2
	2033	3.1	0.9
25 Sa	0233	4.8	1.5
	0925	0.1	0.0
	1722	4.0	1.2
	2147	3.2	1.0
26 Su	0328	4.7	1.4
	1025	0.2	0.1
	1821	4.2	1.3
	2311	3.1	0.9
27 M	0434	4.5	1.4
	1125	0.1	0.0
	1910	4.4	1.3
28 Tu	0021	2.7	0.8
	0547	4.5	1.4
	1225	0.0	0.0
	1951	4.7	1.4
29 W	0115	2.1	0.6
	0659	4.5	1.4
	1318	0.0	0.0
	2030	4.9	1.5
30 Th	0204	1.6	0.4
	0808	4.6	1.4
	1407	0.0	0.0
	2105	5.3	1.6

MAY

Day	Time h.m.	Height ft.	Height m.
1 F	0253	0.6	0.2
	0911	4.8	1.5
	1453	0.2	0.1
	2141	5.6	1.7
2 Sa	0336	-0.2	-0.1
	1011	4.9	1.5
	1539	0.4	0.1
	2216	5.9	1.8
3 Su	0422	-0.8	-0.2
	1110	5.0	1.5
	1623	0.8	0.2
	2253	6.2	1.9
4 M	0510	-1.3	-0.4
	1207	5.0	1.5
	1709	1.3	0.4
	2335	6.3	1.9
5 Tu	0559	-1.5	-0.5
	1306	4.9	1.5
	1756	1.8	0.5
6 W	0018	6.3	1.9
	0648	-1.6	-0.5
	1405	4.8	1.5
	1850	2.2	0.7
7 Th	0105	6.1	1.9
	0743	-1.3	-0.4
	1510	4.8	1.5
	1949	2.6	0.8
8 F	0155	5.8	1.8
	0839	-1.0	-0.3
	1615	4.7	1.4
	2059	2.8	0.9
9 Sa	0253	5.4	1.6
	0941	-0.6	-0.2
	1722	4.8	1.5
	2223	2.7	0.8
10 Su	0359	5.0	1.5
	1044	-0.2	-0.1
	1821	4.9	1.5
	2344	2.5	0.8
11 M	0511	4.6	1.4
	1147	0.1	0.0
	1913	5.0	1.5
12 Tu	0051	2.0	0.6
	0627	4.3	1.3
	1243	0.3	0.1
	1958	5.2	1.6
13 W	0147	1.5	0.5
	0741	4.2	1.2
	1333	0.6	0.2
	2037	5.3	1.6
14 Th	0236	1.0	0.3
	0844	4.1	1.2
	1417	0.8	0.2
	2109	5.3	1.6
15 F	0318	0.5	0.2
	0940	4.1	1.2
	1459	1.1	0.3
	2138	5.4	1.6
16 Sa	0353	0.1	0.0
	1031	4.2	1.3
	1535	1.4	0.4
	2205	5.4	1.6
17 Su	0427	-0.2	-0.1
	1121	4.2	1.3
	1610	1.8	0.5
	2233	5.4	1.6
18 M	0500	-0.4	-0.1
	1203	4.2	1.3
	1645	2.1	0.6
	2259	5.4	1.6
19 Tu	0532	-0.5	-0.2
	1249	4.2	1.3
	1720	2.4	0.7
	2327	5.4	1.6
20 W	0604	-0.6	-0.2
	1331	4.2	1.3
	1756	2.7	0.8
	2358	5.4	1.6
21 Th	0643	-0.6	-0.2
	1416	4.2	1.3
	1831	3.0	0.9
22 F	0033	5.3	1.6
	0721	-0.5	-0.2
	1502	4.3	1.3
	1916	3.2	1.0
23 Sa	0112	5.2	1.6
	0803	-0.4	-0.1
	1554	4.3	1.3
	2013	3.3	1.0
24 Su	0200	4.9	1.5
	0851	-0.3	-0.1
	1643	4.4	1.3
	2122	3.2	1.0
25 M	0253	4.7	1.4
	0944	-0.2	-0.1
	1735	4.6	1.4
	2240	2.9	0.9
26 Tu	0358	4.4	1.3
	1040	0.0	0.0
	1819	4.8	1.5
	2351	2.4	0.7
27 W	0515	4.2	1.3
	1137	0.2	0.1
	1859	5.1	1.6
28 Th	0147	1.5	0.5
	0637	4.1	1.2
	1333	0.4	0.1
	1941	5.4	1.6
29 F	0143	0.9	0.3
	0756	4.2	1.3
	1326	0.7	0.2
	2018	5.8	1.8
30 Sa	0234	0.0	0.0
	0908	4.3	1.3
	1420	1.0	0.3
	2100	6.1	1.9
31 Su	0321	-0.8	-0.2
	1014	4.5	1.4
	1509	1.3	0.4
	2139	6.4	2.0

JUNE

Day	Time h.m.	Height ft.	Height m.
1 M	0410	-1.4	-0.4
	1114	4.7	1.4
	1556	1.7	0.5
	2222	6.6	2.0
2 Tu	0457	-1.7	-0.5
	1211	4.9	1.5
	1647	2.0	0.6
	2307	6.7	2.0
3 W	0546	-1.9	-0.6
	1307	4.9	1.5
	1739	2.3	0.7
	2354	6.6	2.0
4 Th	0636	-1.8	-0.5
	1402	5.0	1.5
	1834	2.6	0.8
5 F	0040	6.3	1.9
	0724	-1.5	-0.5
	1455	5.0	1.5
	1934	2.7	0.8
6 Sa	0133	5.8	1.8
	0817	-1.1	-0.3
	1552	5.0	1.5
	2045	2.8	0.9
7 Su	0228	5.3	1.6
	0909	-0.6	-0.2
	1646	5.0	1.5
	2203	2.6	0.8
8 M	0330	4.8	1.5
	1006	-0.1	0.0
	1741	5.1	1.6
	2317	2.3	0.7
9 Tu	0439	4.3	1.3
	1059	0.4	0.1
	1824	5.2	1.6
10 W	0026	1.9	0.6
	0558	3.9	1.2
	1155	0.8	0.2
	1909	5.3	1.6
11 Th	0123	1.4	0.4
	0719	3.7	1.1
	1242	1.2	0.4
	1944	5.4	1.6
12 F	0213	0.9	0.3
	0833	3.7	1.1
	1335	1.5	0.5
	2019	5.5	1.7
13 Sa	0252	0.4	0.1
	0934	3.8	1.2
	1417	1.9	0.6
	2100	5.6	1.7
14 Su	0331	0.0	0.0
	1027	4.0	1.2
	1500	2.2	0.7
	2121	5.7	1.7
15 M	0406	-0.3	-0.1
	1115	4.1	1.2
	1541	2.5	0.8
	2153	5.8	1.8
16 Tu	0441	-0.5	-0.2
	1158	4.3	1.3
	1617	2.7	0.8
	2225	5.8	1.8
17 W	0513	-0.7	-0.2
	1239	4.4	1.3
	1655	2.9	0.9
	2300	5.8	1.8
18 Th	0545	-0.8	-0.2
	1319	4.4	1.3
	1733	3.0	0.9
	2333	5.7	1.7
19 F	0621	-0.8	-0.2
	1401	4.5	1.4
	1812	3.1	0.9
20 Sa	0011	5.6	1.7
	0659	-0.8	-0.2
	1439	4.6	1.4
	1859	3.1	0.9
21 Su	0053	5.4	1.6
	0739	-0.7	-0.3
	1518	4.7	1.4
	1952	3.1	0.9
22 M	0135	5.1	1.6
	0824	-0.5	-0.2
	1602	4.8	1.5
	2055	2.9	0.9
23 Tu	0231	4.8	1.5
	0910	-0.2	-0.1
	1644	5.0	1.5
	2206	2.6	0.8
24 W	0337	4.4	1.3
	1002	2.2	0.7
	1728	5.3	1.6
	2320	2.0	0.6
25 Th	0457	4.1	1.2
	1058	0.6	0.2
	1811	5.6	1.7
26 F	0025	1.3	0.4
	0628	3.9	1.2
	1153	1.0	0.3
	1856	5.9	1.8
27 Sa	0123	0.5	0.2
	0754	4.0	1.2
	1253	1.5	0.5
	1940	6.3	1.9
28 Su	0216	-0.3	-0.1
	0913	4.2	1.3
	1351	1.8	0.5
	2026	6.6	2.0
29 M	0308	-0.9	-0.3
	1015	4.5	1.4
	1447	2.1	0.6
	2112	6.8	2.1
30 Tu	0357	-1.4	-0.5
	1113	4.8	1.5
	1540	2.3	0.7
	2201	6.9	2.1

Tides

TYPICAL TIDE CURVES FOR UNITED STATES PORTS

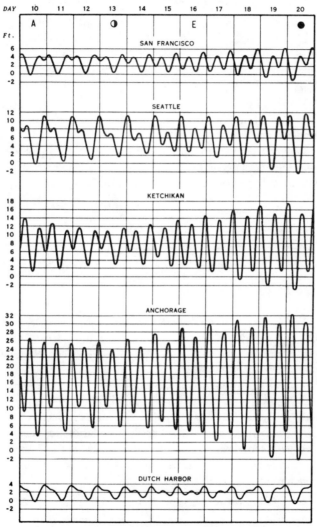

A discussion of these curves is given on the preceding page.

Lunar data: A – Moon in apogee
                ◑ – last quarter
                E – Moon on Equator
                ● – new Moon

Photographing the Moon

The Moon is the biggest and brightest object in the night sky, and attracts the interest of many photographers. Even when the moon is near the horizon, however, and seems to be quite large, it is difficult to capture on film. With a standard 50mm lens, the image on a negative or slide will be less than 1/32 inch in diameter (about .45 mm). Even if the image is enlarged 15 times, it will still be only about 1/4 inch in diameter (about 7 mm).

Longer lens will give better results, but use a tripod with long lens because of the longer exposure times needed. To determine the final size of the moon, use this formula:

$$\text{image size on film} = \frac{\text{focal length}}{110}$$

For example:

Lens Size	Image Size
100mm	0.91 mm
200mm	1.8 mm
300mm	2.7 mm

You can also use converters to improve the power of your lens. A x2 teleconverter will double the focal length of a telephoto lens. However, the f-ratio is also doubled, requiring longer exposures. For example, a 100mm lens at f/8 becomes 200mm at f/16 using a x2 teleconverter.

Exposure time for lunar photos is an important consideration, because the earth is rotating while the shutter is open. Clock drives that synchronize telescopes with stellar motions can be used with some cameras to minimize this motion. Without a clock drive, maximum exposure time for a fixed camera can be determined using this formula:

$$\text{exposure (in seconds)} = \frac{250}{F \text{ (focal length in mm)}}$$

The following formula allows you to calculate the exposure time needed for different types of film and different phases of the moon. The film speed is indicated by A (ISO number), f is the f-stop the lens is open to, and B represents a value corresponding to the brightness of the moon (10 for thin crescent, 20 for wide crescent, 40 for quarter moon, 80 for gibbous moon, 200 for full moon).

$$\text{time (in seconds)} = \frac{f^2}{(A \times B)}$$

Rules of Thumb for Photographing the Moon

- Always use a tripod.
- Photos of the full moon are flat and featureless. For more interesting pictures, photograph the moon at crescent or quarter phases when the mountains and craters are illuminated from the side and cast shadows.
- Always bracket exposures since exposure times given by formulas are approximate, varying according to the exact phase of the moon, atmospheric conditions, etc. To be safe, bracket at least one and preferably two stops on either side of the exposure suggested by the formula.

Lunar Photography with a Telescope

There are several methods for photographing the moon using a telescope. With any of these methods, it is fairly simple compared to photographing other astronomical objects, and almost any kind of film will give satisfactory results. The main problems is the great contrast range in the image, from bright sunlight to deep shadow. The exception is the full moon, which is very bright with little contrast. The best film is one that has great exposure latitude—it can tolerate considerable over- or under-exposure. Mistakes can then be compensated for in developing and printing, whether you do it yourself or have it done by a lab (black and white films are the best to use for this). It is best to choose one or two films and stick with them so that you can learn their qualities. Some standard films are: Kodak Tri-X Pan (400 ASA), Plus-X (125 ASA), T-Max 100, and T-Max 400.

Color films have much less latitude than black and white. If you can make your own prints, use color negative films and manipulate them in the darkroom. Any of the Kodacolor VR or the new Kodacolor VRG print films are good. Color reversal (slide) films have almost no exposure latitude, so bracketing is particularly important. Kodachrome 64, Ektachrome 64, and Ektachrome 400 are the most common and are also easily available. If you bracket carefully (two stops under and one stop over with slide film), you are almost certain to get a few excellent shots from a roll.

■ Prime focus method. This is the simplest method for lunar photography. If you have a catadioptic telescope, such as a Schmidt-Cassegrain or Maksutov, remove the camera lens and using a simple adaptor (T-ring, available at camera stores and astronomical supply houses), attach the camera directly to the telescope (without its eyepiece). The telescope becomes the camera lens, in effect, a very long telephoto lens. For example, an 8-inch Celestron or Meade Schmidt-Cassegrain telescope becomes a 1250mm, f/10 telephoto.

■ Afocal method. This method requires two tripods, one for the camera, and one for the telescope. The camera (with lens) is set up as close to the telescope eyepiece as possible, and focused on the image seen through the camera. The exposure time is calculated by determining the focal length (F) and the f/ratio of the telescope system.

F = focal length of camera lens x magnification of telescope

f/ratio = F/diameter of the telescope objective

Magnification of the telescope = $\dfrac{\text{focal length of telescope}}{\text{focal length of eyepiece}}$

■ Projection method. This method requires the use of a special adaptor which is much more expensive than the simple T-ring used in the prime focus method. The adaptor connects the camera without its lens to the eyepiece of the telescope. This is an advanced method used in most amateur astrophotography of planets, double stars, and deep-sky objects involving very long exposures and special films. It can also be used to photograph the moon.

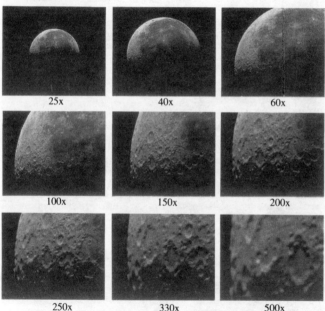

25x   40x   60x

100x   150x   200x

250x   330x   500x

Comparison of telescope magnifications
(Photograph provided courtesy of Celestron International.)

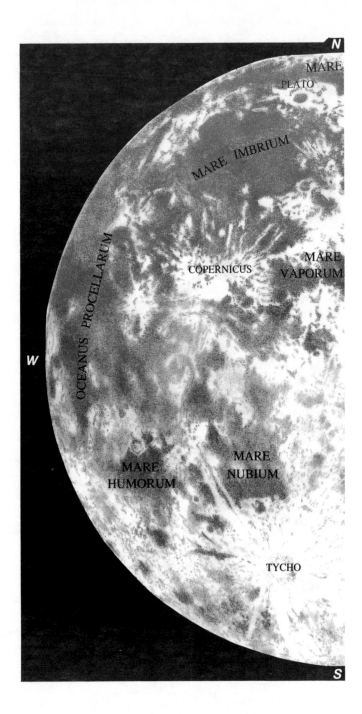

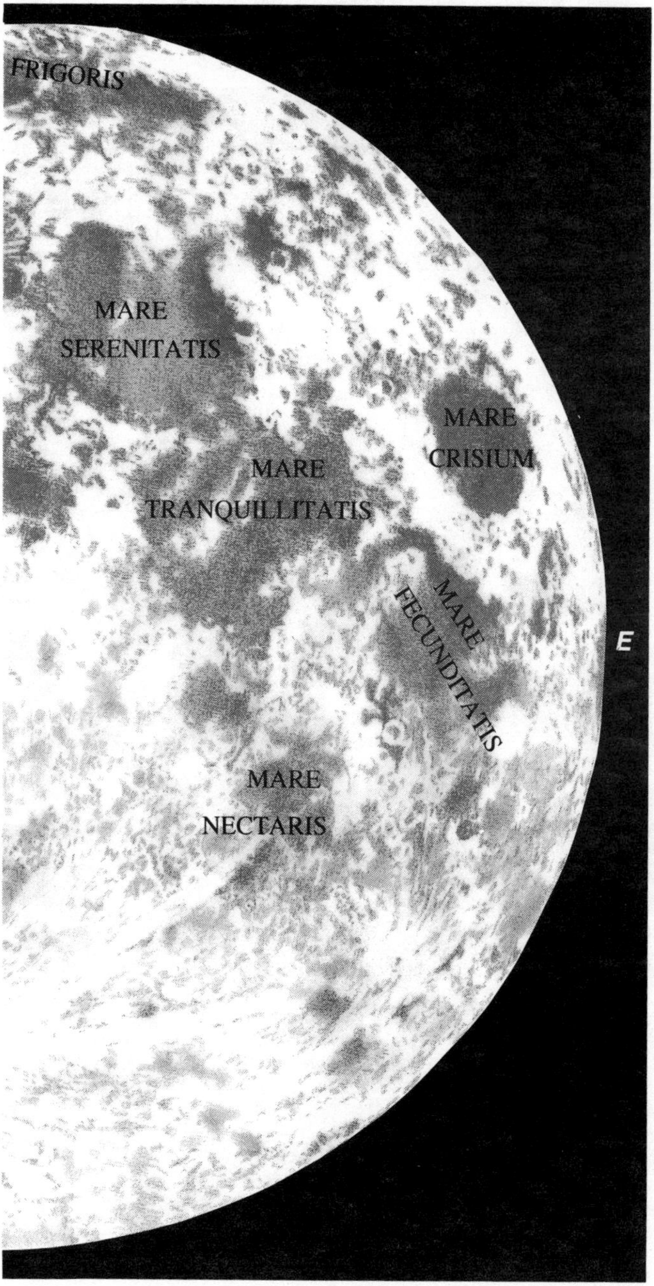

FRIGORIS

MARE
SERENITATIS

MARE
CRISIUM

MARE
TRANQUILLITATIS

MARE
FECUNDITATIS

E

MARE
NECTARIS

Some telescopes (Newtonians) invert images from top to bottom.

Some telescopes (refractors and catadioptics with star diaginals) reverse images from left to right.

South Polar Region
(Stereographic projection of polar areas from 45 deg. to 90 deg.)

Maps on pages 72-79 are prepared and published by the Defense Mapping Agency Aerospace Center (St. Louis, Missouri) for NASA.

North Polar Region

Maps on pages 74-79 are a mercator projection of lunar areas from 45° North to 45° South.
Scale 1:10,000,000 at 34° North and South latitudes.

Heavy vertical lines on pages 75 and 78 mark the edge of the visible (near) face of the moon.

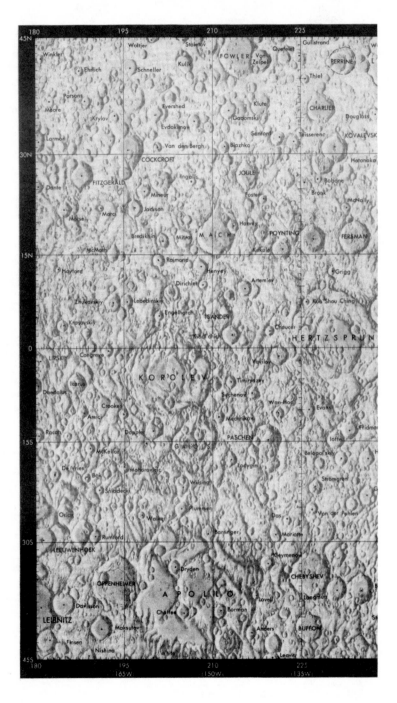

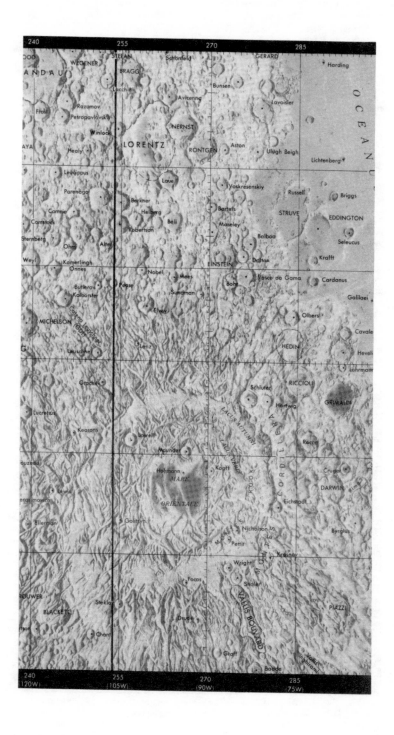

240 255 270 285

WEGENER · Schönfeld · GERARD · Harding
ANDAU · STEFAN · BRAGG · Bunsen · O
· Lacchini · Lavoisler C
Frost · Razumov · Avicenna E
· Petropavlovskiy · NERNST A
· Winlock · LORENTZ · RONTGEN · Aston · Ulugh Beigh N
AYA · Healy · Lichtenberg U
· Leucippus · Laue · Voskresenskiy · Russell · Briggs
· Parenago · Barkner · Bartels · STRUVE · EDDINGTON
· Camsic · Helberg · Bell · Moseley · Balboa · Seleucus
· Carnicod · Robertson · Krafft
Sternberg · Ohm · Alter · EINSTEIN · Dalton
Weyl · Kamerlingh · Nobel · Mees · Vasco da Gama · Cardanus
Onnes · Repse · Sundman · Bohr
· Butlerov · Elvey · Olbers · Galilaei
· Kolhorster
MICHELSON · Cavale
· Lauschner · Lenz · HEDIN · Heveli
G · Lohrman
· Gratchev · Schlüter · RICCIOLI
· Lucretius · Hartwig · GRIMALDI
· Kearons · Lowell · Kopff · Recca
· Maunder · LACUS AUTUMNI · Crüger
· Hohmann · MARE · Kopff · DARWIN
· Lewis · ORIENTALE · Eichstadt · Byrgius
· Ellerman · Golitsyn · Nicholson · Pettit · Krasnov
· Wright · Shaler
ROUWER · Steklov · Focas · PIAZZI
BLACKETT · VALLIS BOUVARD
· Ghant · Dale · Graff · highlands
· Baade

240 255 270 285
(120W) (105W) (90W) (75W)

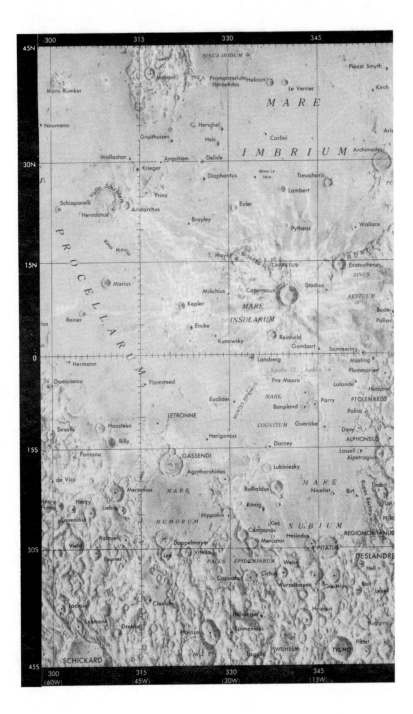

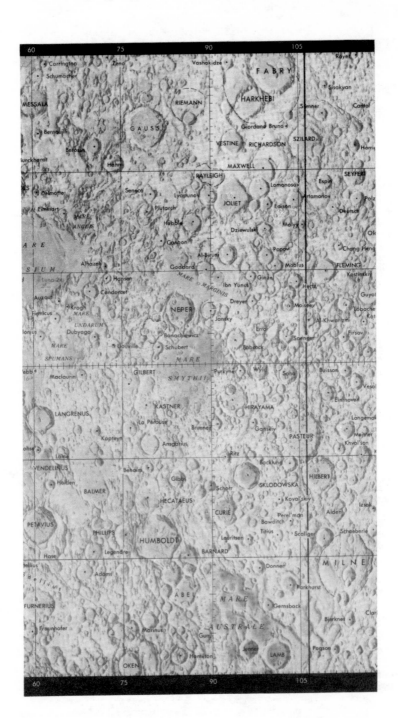

Royer
· Carrington　　Zeno　　　　Vashakidze
· Schumacher　　　　　　　　　　　F A B R Y
　　　　　　　　　　　　　　　　　　　　　Sisakyan
MESSALA　　　　　　RIEMANN　　HARKHEBI　　　　　　Cantol
　　　　　　　　　　　　　　　　　Giordano Bruno
· Bernouilli　　　G A U S S　　　　　　　　　　SZILARD
Berosu　　　　　　　　　　VESTINE　RICHARDSON　　　　Harri
lurckhardt　　　Hahn　　　　　　　MAXWELL
　　　　　　　　　　　RAYLEIGH　　　　　　　　　　　SEYFERT
· Delmotte　　Seneca　Lyapunov　　Lomonosov　Espin
· Eimmart　　　　Plutarch　JOLIET　　Edison　Artamonov　Polz
MARE　　　　　　　Hubble　　　　　　　Malyx　Deutsch
ANGUE　　　　　　　　　Dziewulski　　　　　　　　Chang Heng
A R E　　　　　· Cannon　　　　　　　Popov　　　Ol
· Alhazen　　　Al-Biruni　　　　　　　　FLEMING
S I U M　　　Goddard　　　　　　Mobius　　Kostinskiy
　　　Lyno-zu　　Hansen　MARE MARGINIS　Ginzel　Hertz　　Guyot
Auzout　　　Condorcet　　Ibn Yunus　　　　　Lobache
Firmicus　· Krogh　MARE　NEPER　Dreyer　　Moiseev　Al-Khwarizmi　Kot
lonius　Dubyago UNDARUM　　Jansky　　Erro　　　Firsov
MARE　Gouville　Banachiewicz　　　Babcock　Saenger
SPUMANS　　Schubert　　　　　　　　　　　Buisson
ebb　· Maclaurin　GILBERT　M A R E　Pyrkyne　Wyld　Saha　　Vesal
　　　　　　　　S M Y T H I I　　　　HIRAYAMA　Einthoven
LANGRENUS　　KASTNER　　　　　　　　　　Langema
　　　　　　La Perouse　　　　Gansky　PASTEUR　Meitner
· Kapteyn　Ansgarius　Brunner　　　　　　Khvol'son
ohse　　Lame　　　　　　Ritz　　Backlund　HILBERT
VENDELINUS　Behaim　　Gibbs　　Schorr　SKLODOWSKA　Izsak
· Holden　　　　　　　　　· Koval'skiy
BALMER　　HECATAEUS　CURIE　Perel'man　Alden　Scheeberle
PETAVIUS　　　　　　　　Bawditch　Titius　Scaliger
· Phillips　HUMBOLDT　Lauritsen　　　　M I L N E
· Hase　Legendre　BARNARD　　　Donner
ellius　· Adams　　　　　　　　Parkhurst
Faellius　A B E L　M A R E　Gernsback
FURNERIUS　　　　　A U S T R A L E　Bjerknes　Clar
· Fraunhofer　Marinus　Gum　　Jenner　LAMB　Pogson
OKEN　　　　Hamilton

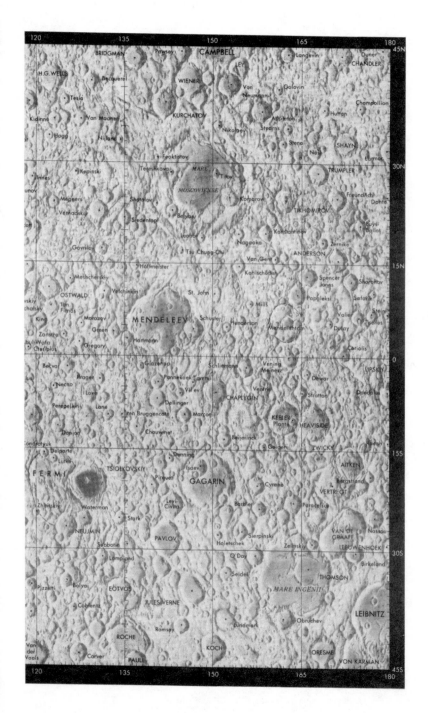

Lunar Gazeteer

Albategnius (crater diameter: 83 miles) 12° South, 4° East
Aliacensis (crater diameter: 50 miles) 31° South, 5° East
Alphonsus (crater diameter: 73 miles) 13° South, 3° West
Anaxagoras (ray crater; crater diameter 32 miles; ray pattern diameter 600 miles) 75° North, 10° West
Anaximander (crater diameter: 55 miles) 66° North, 48° West
Anaximines (crater diameter: 49 miles) 75° North, 45° West
Archimedes (crater diameter: 50 miles) 30° North, 4° West
Aristillus (ray crater; crater diameter: 35 miles; ray pattern diameter: 400 miles) 34° North, 1° East
Aristoteles (crater diameter: 55 miles) 50° North, 18° East
Arzachel (crater diameter: 61 miles) 18° South, 2° West
Atlas (crater diameter: 54 miles) 47° North, 44° East

Bailly (crater diameter: 184 miles) 66° South, 65° East
Barocius (crater diameter: 54 miles) 45° South, 17° East
Berosus (crater diameter: 45 miles) 33° North, 70° East
Blancanus (crater diameter: 72 miles) 64° South, 21° West
Boguslawsky (crater diameter: 61 miles) 75° South, 45° East
Byrgius (crater diameter: 51 miles) 25° South, 65° West

Casatus (crater diameter: 59 miles) 75° South, 35° West
Catherina (crater diameter: 63 miles) 18° South, 24° East
Clairaut (crater diameter: 47 miles) 48° South, 14° East
Clavius (crater diameter: 144 miles) 58° South, 14° West
Colombo (crater diameter: 46 miles) 15° South, 46° East
Condorcet (crater diameter: 49 miles) 12° North, 70° East
Copernicus (ray crater; crater diameter: 57 miles; ray pattern diameter: 750 miles) 10° North, 20° West
Cuvier (crater diameter: 48 miles) 50° South, 10° East
Cyrillus (crater diameter: 58 miles) 13° South, 24° East

Endymion (crater diameter: 77 miles) 55° North, 55° East

Fabricus (crater diameter: 48 miles) 43° South, 42° East
Faraday (crater diameter: 45 miles) 42° South, 8° East
Fracastorius (crater diameter: 75 miles) 21° South, 33° East
Furnerius (crater diameter: 81 miles) 36° South, 60° East

Gassendi (crater diameter: 69 miles) 18° South, 40° West
Gemma Frisius (crater diameter: 55 miles) 34° South, 14° East

Geminus (crater diameter: 54 miles) 35° North, 57° East
Grimaldi (crater diameter: 127 miles) 6° South, 68° West
Gruemberger (crater diameter: 58 miles) 68° South, 10° West

Hainzel (crater diameter: 46 miles) 41° South, 34° West
Hevel (crater diameter: 69 miles) 2° North, 67° West
Hipparchus (crater diameter: 95 miles) 6° South, 5° East
Hommel (crater diameter: 75 miles) 54° South, 33° East
Humboldt (crater diameter: 130 miles) 27° South, 81° East

Inghirami (crater diameter: 57 miles) 48° South, 70° West

Kepler (ray crater; crater diameter: 20 miles; ray pattern diameter:
    400 miles) 8° North, 38° West
Kircher (crater diameter: 48 miles) 67° South, 45° West

Lacus Somniorum 37° North, 35° East
Lacus Mortis 44° North, 27° East
Langrenus (ray crater; crater diameter: 82 miles; ray pattern diamter:
    950 miles) 9° South, 61° East
Legendre (crater diameter: 46 miles) 29° South, 70° East
Letronne (crater diameter: 73 miles) 10° South, 43° West
Licetus (crater diameter: 47 miles) 47° South, 6° East
Longomontanus (crater diameter: 92 miles) 50° South, 21° West

Maginus (crater diameter: 116 miles) 50° South, 6° West
Manzinus (crater diameter: 60 miles) 68° South, 25° East
Mare Anguis 23° North, 69° East
Mare Australe 50° South, 80° East
Mare Crisium 18° North, 58° East
Mare Fecunditatis 4° South, 51° East
Mare Frigoris 55° North, 0
Mare Humboldtianum 55° North, 75° East
Mare Humorum 23° South, 38° West
Mare Imbrium 36° North, 16° West
Mare Marginis 13° North, 87° East
Mare Nectaris 14° South, 34° East
Mare Nubium 19° South, 14° West
Mare Orientale 19° South, 95° West
Mare Serenitatis 30° North, 17° East
Mare Smythii 3° South, 80° East

Mare Spumans 1° North, 65° East
Mare Tranquillitatis 9° North, 30° East
Mare Vaporum 14° North, 5° East
Maurolycus (crater diameter: 72 miles) 42° South, 14° East
Mersenius (crater diameter: 51 miles) 21° South, 49° West
Metius (crater diameter: 54 miles) 40° South, 44° East
Mons La Hire 28° North, 25° West
Mons Piton 41° North, 1° West
Montes Apenninus 20° North, 2° West
Montes Carpatus 15° North, 24° West
Montes Caucasus 36° North, 8° East
Montes Cordillera 27° North, 85° West
Montes Haemus 16° North, 14° East
Montes Jura 46° North, 38° West
Montes Pyrenaei 14° South, 41° East
Montes Riphaeus 6° South, 26° West
Montes Taurus 28° North, 35° East
Moretus (crater diameter: 73 miles) 70° South, 8° West
Mutus (crater diameter: 47 miles) 63° South, 30° East

Neper (crater diameter: 75 miles) 7° North, 83° East
Newton (crater diameter: 85 miles) 78° South, 20° West

Oceanus Procellarum 10° North, 47° West
Olbers (ray crater; crater diameter: 42 miles; ray pattern diameter: 500 miles) 7° North, 78° West
Orontius (crater diameter: 74 miles) 40° South, 4° West

Palus Epidemiarum 31° South, 26° West
Palus Nebularum 38° North, 1° East
Palus Putredinis 27° North, 1° West
Palus Somnii 15° North, 46° East
Petavius (crater diameter: 110 miles) 25° South, 61° East
Phocylides (crater diameter: 75 miles) 54° South, 58° West
Piccolomini (crater diameter: 56 miles) 30° South, 32° East
Pitiscus (crater diameter: 51 miles) 51° South, 31° East
Plato (crater diameter: 63 miles) 51° North, 9° West
Pontecoulant (crater diameter: 60 miles) 69° South, 65° East
Posidonius (crater diameter: 63 miles) 32° North, 30° East
Proclus (ray crater; crater diameter: 19 miles; ray pattern diameter: 400 miles) 16° North, 47° East
Ptolemaeus (crater diameter: 93 miles) 14° South, 3° West
Purbach (crater diameter: 77 miles) 25° South, 2° West

Pythagoras (crater diameter: 80 miles) 65° North, 65° West

Riccioli (crater diameter: 99 miles) 3° South, 75° West
Rosenberger (crater diameter: 61 miles) 55° South, 43° East
Rupes Philolaus 68° North, 25° West
Rupes Recta 22° South, 8° West

Sacrobosco (crater diameter: 60 miles) 24° South, 17° East
Scheiner (crater diameter: 71 miles) 60° South, 28° West
Schickard (crater diameter: 134 miles) 44° South, 54° West
Schiller (crater diameter: 112 miles) 52° South, 39° West
Schomberger (crater diameter: 52 miles) 76° South, 30° East
Sinus Iridum 45° North, 32° West
Sinus Aestuum 12° North, 9° West
Sinus Medii 0°, 0°
Sinus Roris 54° South, 46° West
Snellius (crater diameter: 50 miles) 29° South, 56° East
Stevinus (crater diameter: 46 miles) 33° South, 54° East
Stoflerus (crater diameter: 84 miles) 41° South, 6° East
Strabo (ray crater; crater diameter: 34 miles; ray pattern diamter: 400
    miles) 62° North, 55° East

Theophilus (ray crater; crater diameter: 64 miles; ray pattern diameter:
    675 miles) 12° South, 26° East
Tycho (ray crater; crater diameter: 54 miles; ray pattern diameter:
    1900 miles) 43° South, 11° West

Vallis Alpes 49° North, 2° East
Vallis Rheita 40° South, 48° East
Vendelinus (crater diameter: 94 miles) 16° South, 62° East
Vieta (crater diameter: 53 miles) 29° South, 57° West
Vlacq (crater diameter: 56 miles) 53° South, 39° East

Walter (crater diameter: 82 miles) 33° South, 1° East
Wargentin (crater diameter: 53 miles) 50° South, 60° West
Wilhelm I (crater diameter: 64 miles) 43° South, 20° West
Wurzelbauer (crater diameter: 54 miles) 34° South, 16° West

Lunar Feature Terms

mare	lava plain	*rima*	rille
mons	mountain	*rupes*	fault
montes	mountain range	*sinus*	bay
palus	uneven dark plain	*vallis*	valley

Early Moon Watchers

The moon, as the most visible symbol in the sky of the measured passage of time, has long been the object of study by humans. Early cultures, however, lacking an understanding of the heliocentric (sun-centered) nature of the solar system, generally confined their studies to support for religious beliefs. The first real development in scientific reasoning came from Nicolas Copernicus (1473 to 1543), who created a theory about the revolution of the moon around the earth and the planets around the sun.

Tycho Brahe, a Danish astronomer (1546 to 1601), believed in a geocentric theory (earth-centered) of the universe, but was responsible for developing accurate measurements of the lunar orbit, including the slight variations created from the effect of the sun's gravity. Johannes Kepler (1571 to 1630) believed, like Copernicus, that the sun was the center of the solar system, and made pertinent observations on the moon's orbit, the lunar effect on tides, and the nature of the lunar surface.

Kepler and Galileo Galilei (1564 to 1642) both contributed to early lunar research by creating and refining telescopic observation of the moon. Along with the widespread adoption of the telescope by astronomers and scientists in the 17th century came the first detailed lunar maps. Johannes Hevelius (1611 to 1687) was an astronomer who specialized in the study of the lunar surface, ultimately publishing a description of the moon which included the first standardized names for lunar features. The Hevelius nomenclature has not survived the test of time, being replaced by that of another astronomer from that era, Giovanni Riccioli (1598 to 1671). Riccioli's lists of lunar features included names of prominent landmarks, craters, and seas, many of which are still in use.

The first map of the moon to use a system of coordinates was created by Johann Mayer (1723 to 1762) in 1750. The first lunar landscapes that included measurements were published in 1791 by Johann Schroeter (1745 to 1816). In 1835,

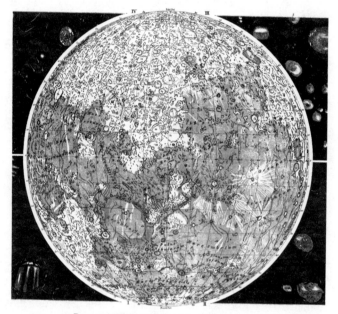

Beer and Madlers Map of the Moon (circa 1851).

exaggerated reports of the astronomer Sir John Herschel (1792 to 1871) and his discoveries about the moon were published in a popular newspaper in New York City. These stories included supposed sightings of lunar animals, but were written without the knowledge or consent of Herschel.

Many other astronomers contributed to the growing study of the moon, taking advantage of developments in technology and science. The first photograph of the moon was produced on March 23, 1840, by John Draper (1811 to 1882). By the end of the 19th century, several major observatories published books of photographs of the moon, including Lick Observatory and the Paris Observatory.

The invention and perfection of rocket-powered flight in the early part of the 20th century was the first step towards a close-up examination of the moon. Although the scientific theory about space flight to the moon was already well developed by the 1950s, it took a U.S.-U.S.S.R. "space race" to create the final impetus for the first unpiloted and piloted expeditions to get there.

Hundreds of pounds of moon rocks, close-up human study, and huge quantities of remote sensing data have resulted from the exploration of the moon. This information has proven useful in developing new theories about the solar system and creation of the planets, and has also proven the potential of the moon for providing huge quantities of valuable materials useful in the further exploration of space.

Moon Rocks

Examination of the rocks brought back by the Apollo astronauts is still underway. Scientists have discovered many interesting features about the composition and origin of the moon from these rocks, most of them formed from cooling lava and therefore igneous in nature. Some of these rocks are similar to the basalt which is found on earth. Samples of this basalt were collected in low areas of the lunar surface that are observed as maria from the earth.

Rocks from higher regions of the moon are also igneous, and are referred to as gabbro, norite, and anorthosite, similar to rocks of the same names on earth. Although moon rocks have some characteristics similar to earth rocks, they are recognizably different because of the complete lack of water and the effects that water has on minerals in rocks. Moon rocks also exhibit crystals of metallic iron that occur because of the lack of free oxygen. Lunar minerals include feldspar, olivine, pyroxine, ilmenite, plagioclase, and troilite.

How old is the moon? The best guess from studying the age of lunar rocks is that the moon was formed about 4.6 billion

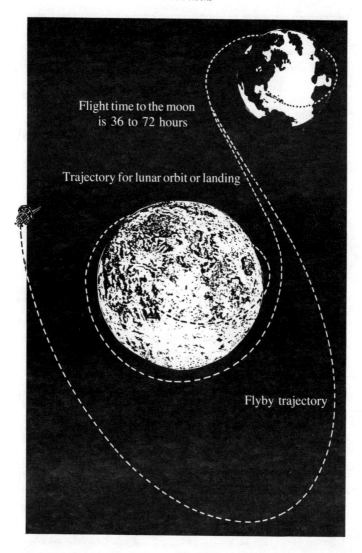

Flight time to the moon
is 36 to 72 hours

Trajectory for lunar orbit or landing

Flyby trajectory

years ago. The newest rocks found so far on the moon (from the dark maria regions) have been dated at 3.1 to 3.8 billion years old.

Estimated Composition of the Moon

Oxide	Moon	Earth's Crust
SiO_2	43.3-48.7%	45.0%
MgO	25.3-33.4	6.8
FeO	11.3-13.9	6.6
Al_2O_3	3.7-7.6	24.6
CaO	3.4-6.1	15.8
Cr_2O_3	0.3-0.4	0.1
TiO_2	0.2-0.4	0.6
Na_2O	0.005-0.15	0.5
K_2O	0.01	0.1

(from *The Geology of the Terrestrial Planets*, NASA, 1984)

Analysis of the moon's composition from rock samples and the investigations of unpiloted probes reveal that it includes: oxygen (up to 42 percent of the content of lunar material), iron, aluminum, carbon, nitrogen, sulfur, calcium, titanium, hafnium, zirconium, magnesium, sodium, oxygen, scandium, yttrium.

Lunar Soil

Material on the surface of the moon is referred to as lunar soil, but it has no organic content. Lunar soil forms a layer from 3 to 60 feet (1 to 20 meters) deep on the surface. This layer is composed of rocks and powder but was not formed by eroding forces such as wind or water. Instead, lunar soil was created over a period of billions of years by the continuous bombardment of meteorites. Larger meteorites form visible craters that can be seen from the earth; smaller, virtually invisible craters are formed by particles of cosmic dust. The smallest craters are only 1/25,000 inch (1/1000 millimeter) in diameter.

Moonquakes

Much is still unknown about the interior of the moon, but scientists believe that there is a core of molten or partly molten material. There are different layers in the structure of the moon, and a constant shifting produces tremors similar to earthquakes. These moonquakes are usually very weak—many of them release no more energy than a firecracker.

Some moonquakes are caused by the impact of meteorites on the surface. Others occur at regular intervals during a lunar cycle, suggesting that gravitational forces from the earth similar to ocean tides are causing movement within the body.

The Creation of the Moon

It is believed that the moon was formed from molten material at least 4.6 billion years ago during the creation of the solar system. One theory suggests that it was formed near to, but separate from the earth; another suggests that it was formed some distance away but was pulled into orbit around the earth by gravitational attraction. During the first phase of the moon's existence, violent forces caused by the cooling of the molten material produced many of the existing features. The largest visible features on the surface were created by the collision with asteroids that may have been as large as the state of Delaware. The interior of the moon is believed to have originally been more solid, with melting caused by radioactive decay. The last violent surface activity on the moon probably happened more than 3 billion years ago, with volcanic-like eruptions and flooding of lava across the surface.

Lagrange Points

The gravitational physics of orbiting bodies produces a unique condition. Where there are examples of two large celestial bodies orbiting in relation to each other, as in the case of the earth orbiting around the Sun, five specific points in the orbital patterns have the effect of cancelling the gravitational and centrifugal pull of the bodies. These points are called Lagrange points, after their discovery by Joseph Louis Lagrange, a French mathematician, in 1772.

Lagrange points are potentially important spots for the future, because spacecraft, space stations, or permanent space colonies could remain in stable orbits at these locations without the need for constant fuel expenditure to maintain position.

The Lagrange points in the earth-moon system are also affected by additional forces from the Sun. In order to remain unaffected by these forces, objects to be placed at the two most stable points, L-4 and L-5, would have to be placed in their own orbits around the central point defined by L-4 or L-5. These orbits would have a radius of about 90,000 miles from the center and be elliptical.

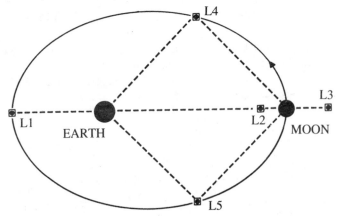

Earth-Moon Lagrange Points

Unpiloted Moon Missions

Spacecraft	Date	Mission-Accomplishments
Pioneer 1 (USA)	October 11, 1958	flyby at 71,300 miles
Pioneer 3 (USA)	December 6, 1958	flyby at 66,654 miles
Luna 1 (USSR)	September 12, 1959	landed on lunar surface
Ranger 6 (USA)	February, 1964	cameras malfunctioned
Ranger 7 (USA)	July 28, 1964	first pre-impact photos

■ *(Mare Nubium, lat 10.6° N, long 24.7 W)*

Ranger 8 (USA)	February 17, 1965	transmitted photographs

■ *Mare Tranquillitatis, lat 2.6° N, long 24.7° W)*

Ranger 9 (USA)	March 21, 1965	transmitted photographs

■ *(interior of Alphonsus, lat 12.9° S, long 2.4° W)*

Luna 5 (USSR)	May, 1965	unsuccessful soft landing
Zond 3 (USSR)	July 18, 1965	orbited, transmitted photos

■*(far side and western limb, altitude 9,960 km to 11,570 km)*

Luna 7 (USSR)	October, 1965	unsuccessful soft landing
Luna 8 (USSR)	December, 1965	unsuccessful soft landing
Luna 9 (USSR)	January 31, 1966	landed, transmitted photos

■*(Oceanus Procellarum, lat 7.1° N, long 65.4° W)*

Luna 10 (USSR)	March 31, 1966	orbited, gamma-ray sensing

■ *(perilune 350 km, apolune 1,015 km)*

Surveyor 1 (USA)	May 30, 1966	landed, transmitted first color photos

■ *(Flamsteed P, lat 2.5° S, long 43.2° W)*

Lunar Orbiter (USA)	August 10, 1966	orbited, transmitted photos, data

■ *(inclination 12°, perilune 190-40km, apolune 1,865-1,815 km)*

Luna 11 (USSR)	August 24, 1966	orbited

■ *(perilune 165 km, apolune 1,195 km)*

Surveyor 2 (USA)	September, 1966	unsuccessful soft landing
Luna 12 (USSR)	October 22, 1966	orbited, transmitted data

■ *(perilune 100 km, apolune 1,740 km)*

Lunar Orbiter 2 (USA)	November 7, 1966	orbited, transmitted photos

■ *(inclination 12°, perilune 50 km, apolune 1,855 km)*

Luna 13 (USSR)	December 21, 1966	landed, transmitted photos, used mechanical soil sampler

■ *(Oceanus Procellarum, lat 18.9° N, long 62.1 W)*

Lunar Orbiter 3 (USA)	February 4, 1967	orbited, transmitted photos

■ *(inclination 21°, perilune 55 km, apolune 1,845 km)*

Surveyor 3 (USA)	April 17, 1967	landed, tested lunar soil transmitted data

■ *(Oceanus Procellarum, lat 3.2° S, long 23.4° W)*

Lunar Orbiter 4 (USA)	May 4, 1967	orbited, remote orbit change

■ *(inclination 85°, perilune 2,705 km, apolune 6,115 km)*

Surveyor 4 (USA)	July 14, 1967	contact lost
Explorer 35 (USA)	July 19, 1967	orbited, magnetic fields

■ *(perilune 830 km, apolune 7,650 km)*   (transmitted data until 2-72)

Spacecraft	Date	Mission-Accomplishments
Lunar Orbiter 5 (USA)	August 2, 1967	orbited, transmitted data
■ *(inclination 85°, perilune 195-100 km, apolune 6,065-1,500 km)*		
Surveyor 5 (USA)	September 8, 1967	landed, soil experiments, transmitted data
■ *(Mare Tranquillitatis, lat 1.4° N, long 23.1° E)*		
Surveyor 6 (USA)	November 7, 1967	landed, transmitted photos, data
■ *(Sinus Medii, lat 0.5° N, long 1.5° W)*		
Surveyor 7 (USA)	January 6, 1968	landed, tested soil, transmitted photos, data
■ *(flank of Tycho, lat 40.9° S, long 11.5° W)*		
Luna 14 (USSR)	April 1968	orbited
Zond 5 (USSR)	September 14, 1968	orbited
Zond 6 (USSR)	November 10, 1968	flyby, returned film to earth
■ *(altitude about 3,300 km)*		
Luna 15 (USSR)	July 1969	landed
Zond 7 (USSR)	August, 1969	flyby, returned film to earth
■ *(western limb and southern far side, altitude about 2,200-10,000 km)*		
Luna 16 (USSR)	September 12, 1970	landed, returned soil samples
■ *(Mare Fecunditatis, lat 0.7° S, long 56.3° E)*		to earth
Zond 8 (USSR)	October 1970	flyby, returned film to earth
■ *(altitude about 1,120 km)*		
Luna 17 (USSR)	November 10, 1970	landed, used remote vehicle (Lunokhod 1), soil tests, transmitted TV signals
■ *(Sinus Iridum, lat 38.3° N, long 35.0° W)*		
Luna 18 (USSR)	September 1971	unsuccessful soft landing
Luna 19 (USSR)	September 28, 1971	orbited, remote sampling
■ *(perilune 140-77 km, apolune 140-385 km)*		
Luna 20 (USSR)	February 1972	landed, returned soil samples
■ *(Crisium basin rim, lat 3.5° N, long 56.5° E)*		to earth
Luna 21 (USSR)	January 8, 1973	landed, remote vehicle, returned soil samples to earth
■ *(Mare fill of Le Monnier, lat 25.8° N, long 30.5° E)*		
Luna 22 (USSR)	May 29, 1974	orbited
Luna 23 (USSR)	November 1974	damage during landing
Luna 24 (USSR)	August 9, 1976	landed, returned with soil samples to earth
■ *(Mare Crisium, lat 12.7° N, long 62.2° E)*		

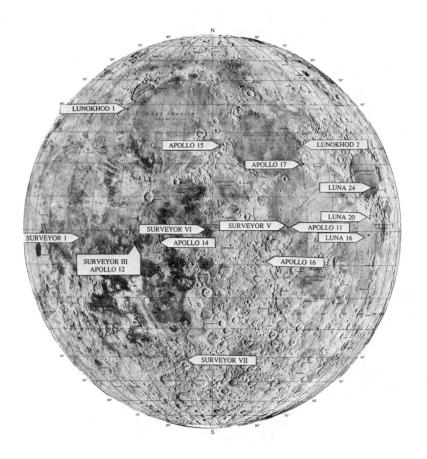

Lunar Landing Site Chart

(Map courtesy of Lunar and Planetary Institute, Houston, Texas)

Apollo Science Experiments

Experiment Number	Experiment	A-11	A-12	A-13	A-14	A-15	A-16	A-17
(Orbital Experiments)								
S-158	Multi-Spectral Photography	•						
S-176	Cm Window Meteoroid					•	•	•
S-177	UV Photography-earth & moon					•	•	
S-178	Gegenschein from Lunar Orbit				•		•	
S-160	Gamma-Ray Spectrometer					•	•	
S-161	X-Ray Fluorescence					•	•	
S-162	Alpha Particle Spectrometer					•	•	
S-164	S-Band Transponder (CSM/LM)				•	•	•	•
S-164	S-Band Transponder (Subsatellite)					•	•	
S-165	Mass Spectrometer					•	•	
S-169	Far UV Spectrometer							•
S-170	Bistatic Radar				•	•	•	
S-171	IR Scanning Radiometer							•
S-173	Particle Shadows/Boundary Layer (Subsatellite)					•	•	
S-174	Magnetometer (Subsatellite)					•	•	
S-209	Lunar Sounder							•
(Surface Experiments)								
S-031	Passive Seismic	•	•		•	•	•	
S-033	Active Seismic				•		•	
S-034	Lunar Surface Magnetometer		•			•		
S-035	Solar Wind Spectrometer		•			•		
S-036	Suprathermal Ion Detector		•		•	•		
S-037	Heat Flow					•	•	•
S-038	Charged Particle Lunar Environment				•			
S-058	Cold Cathode Ion Gauge		•		•	•		
S-059	Lunar Field Geology	•	•		•	•	•	•
S-078	Laser Ranging Retro-Reflector	•			•	•		
S-080	Solar Wind Composition	•	•		•	•	•	
S-151	Cosmic-Ray Detection (Helmets)	•						

Experiment Number (Surface Experiment)	Experiment	Mission A-11	A-12	A-13	A-14	A-15	A-16	A-17
S-152	Cosmic-Ray Detector (Sheets)						•	
S-184	Lunar Surface Closeup Photography	•	•					
S-198	Portable Magnetometer						•	
S-199	Lunar Gravity Traverse							•
S-200	Soil Mechanics					•	•	•
S-201	Far UV Camera/Spectroscope						•	
S-202	Lunar Ejecta and Meteorites							•
S-203	Lunar Siesmic Profiling							•
S-204	Surface Electrical Properties							•
S-205	Lunar Atmospheric Composition							•
S-207	Lunar Surface Gravimeter							•
M-515	Lunar Dust Detector	•	•			•		
S-229	Lunar Neutron Probe							•

Landing Sites

A-11: Sea of Tranquility
A-12: Ocean of Storms
A-13: Mission aborted
A-14: Fra Mauro
A-15: Hadley-Apennine
A-16: Descartes
A-17: Taurus-Littrow

Capsule History of Men on the Moon
- 6 manned missions to lunar surface
- 12 astronauts walk on the moon
- 160 hours spent on the moon
- 59 miles (95 kilometers) traveled on lunar surface
- 2196 samples of lunar material collected—838 pounds (380 kilograms) of lunar rocks and soil

---

Men on the Moon—
The Apollo Program

Mission	Date	Duration	Crew
Apollo 7	October 11, 1968	10d 20h 9m	Cunningham, Eisele, Schirra

■ *orbital tests around earth of Apollo command and service module*

Apollo 8	December 21, 1968	6d 3h 0m	Anders, Borman, Lovell

■ *first flight to moon, first orbit around moon*
(Inclination about 13°, perilune 110 km)

Apollo 9	March 3, 1969	10d 1h 0m	McDivitt, Schweickart, Scott

■ *orbital tests around earth, first flight of complete spacecraft*

Apollo 10	May 18, 1969	8d 0h 3m	Cernan, Stafford, Young

■ *orbital tests around moon of complete Apollo spacecraft, partial descent lunar surface by lunar module*
(Inclination about 1°, perilune 110-15 km)

Apollo 11	July 16, 1969	8d 3h 18m	Aldrin, Armstrong, Collins

■ *first moon landing, first walk on moon (Armstrong)*
(Mare Tranquillitatis (Statio Tranquillitatis), lat 0.7° N, long 23.4° E)

Apollo 12	November 12, 1969	10d 4h 36m	Bean, Conrad, Gordon

■ *lunar landing, surface exploration*
(Oceanus Procellarum, lat 3.2° S, long 23.4° W)

Apollo 13	April 11, 1970	5d 0h 1m	Haise, Lovell, Swigert

■ *flyby, mission aborted in 3rd day*

Apollo 14	January 31, 1971	9d 0h 1m	Mitchell, Roosa, Shepard

■ *lunar landing, surface exploration*
(Fra Mauro highlands, lat 3.7° S, long 17.5° W)

Apollo 15	July 26, 1971	12d 7h 11m	Irwin, Scott, Worden

■ *first use of Lunar Rover, first continuous color television broadcast of moonwalking, extensive study of lunar surface*
(Palus Putredinis (Apenninus-Hadley region), lat 26.1° N, long 3.7° E)

Apollo 16	April 16, 1972	11d 1h 51m	Duke, Mattingly, Young

■ *lunar landing, surface exploration*
(Descartes highlands, lat 9.0° S, long 15.5° E)

Apollo 17	December 7, 1972	12d 13h 1m	Cernan, Evans, Schmitt

■ *lunar landing, first geological study of lunar surface (Schmitt)*
(Taurus-Littrow valley, lat 20.2° N, long 30.8° E)

Moon Calendars

Most ancient civilizations based their calendars on the lunar cycle. The highly visible phase changes of the moon made observations and accurate date projections relatively easy. There is, however, a built-in problem with this method. Most of the seasonal variations in climate are linked to the solar year, and there is no even number of lunar months which equals one solar year. Therefore, calendars based on lunar months are always out of step with the seasons, and extra days or months must be added periodically to make them practical. This procedure is referred to as intercalation.

The ancient Babylonian calendar was based on a lunar month that began when the first crescent moon was visible. There were 12 months in every year, with months alternating between 30 and 29 days long. This lunar calendar would have been out of phase with the solar year, which is almost 12 days longer, except that the Babylonians added special intermediate (or intercalary) months 7 times in every calendar cycle. The extra days in these special months made the cycle come out even (start repeating itself) every 19 years.

The 19-year cycle was also adopted by other cultures, including the Greeks, and is also used in the present-day Jewish calendar. Each of these cycles is the equivalent of 235 lunar months, or lunations.

The Egyptian civilization was heavily influenced by the annual flooding of the Nile River. The time of flooding every year coincided with the rising of the star Sirius in the east, so the Egyptians created a lunar calendar that had its first month begin when the new moon occurred after the rising of Sirius. This lunar calendar featured 12 months, each one 29 1/2 days long. Extra months were included occasionally to keep pace with the solar year. The Egyptians were very advanced in observations of celestial cycles and eventually created a solar calendar to replace the lunar one.

The first era of Greek civilization relied on lunar calendars that were organized and maintained by each separate city or town. During a period that began about the sixth century B.C., Greek astronomers and mathematicians created more organized lunar calendars based on 19-year cycles. The Greeks eventually switched to the Roman calendar, which was based on the Sun.

The Romans originally used a lunar calendar but reorganized it during the reign of Julius Caesar. Caesar added extra days to the lunar calendar to keep it from getting out of step with the solar year; the new calendar was first used in 45 B.C. This calendar was referred to as the Julian calendar and was widely used in western countries until a further reform was organized in 1582. At that time, the rapid spread of Christianity necessitated the fixation of religious holidays based on the celebration of Easter. Pope Gregory XIII was responsible for this reform, and the new calendar, still in use today, is called the Gregorian calendar. The Gregorian calendar measures a year at 365 days (an average of 365.2425 days if leap year is included) and comes very close to matching the solar cycle. Every 400 years, the total difference between the two is only a few hours. By comparison, the error with the Julian calendar was about 3 days for the same period.

The Gregorian calendar which we use today is based on a demand from the Christian religion to determine the correct date for Easter every year. Easter is considered the beginning of the Christian calendar and all Christian holidays and special days are figured from that date. Easter is determined by a fixed set of rules that modify the actual lunar cycle in order to keep the date between March 22 and April 25. If the real, observable lunar cycle were used, the date for Easter could vary much more.

In practice, a fixed date for the spring equinox is set at March 21, and Easter is determined by finding the first Sunday following the full moon that occurs on or after March 21. This full moon date is not determined from the observable cycle but by special religious tables that vary slightly from the real times. Not by coincidence, these tables are based on the same 19-year lunar cycle which was used by the Babylonian, Greek, and

Jewish calendars.

The Jewish calendar was first created around the sighting of the crescent moon, which determined the beginning of each of 12 lunar months. At occasional intervals over the years, the twelfth month was repeated to synchronize the calendar with the seasons. Until the fourth century A.D., there were various versions of this calendar in use. Jews in some countries began their year with the month of Nisan in the spring; other countries used the month of Tishri in the fall. The modern Jewish calendar was adopted in the fourth century A.D. and relies on fixed sequences of months based on a 19-year cycle of lunations. This 19-year cycle was the same one used by the Babylonians.

The new calendar determines the beginning of the year from the time of the new moon in the month of Tishri (occurring in the fall) but uses special rules to keep important religious days from falling on the wrong days. The Jewish year can have 353, 354, or 355 days, with occasional "leap years" of 383, 384, or 385 days. For every 19-year cycle, there are 12 regular years and 7 "leap years."

The Moslem calendar is also based on lunar months but uses a cycle of 33 years. There are 12 lunar months in the Moslem calendar, with lengths alternating between 29 and 30 days. The calendar is fixed instead of relying on actual observation of the moon, but it is only off from the actual lunar cycle by one day every 2,500 years.

In practice, many Moslems base their religious holidays on actual observation but rely on a modified lunar calendar—or western Gregorian calendars—for day-to-day activities. The important dates are figured from the first observation of the crescent moon, and days begin at sunset. The first year of use for the Moslem calendar was 622 A.D., the year Mohammed left Mecca.

Most of India also uses the Gregorian calendar for day-to-day activities, but Hindu religious dates are based on a very old lunar calendar. The traditional Hindu calendar dates back to at least 1000 B.C. It is organized around 12 lunar months of

27 or 28 days each. About every 60 months, an extra month is added to keep it in step with the solar year. Each month is also divided into two parts, corresponding to the waxing and waning phases. In addition, the Hindu calendar also uses a complicated system relating earth's solar orbit with Jupiter's, and uses a unique division of daily time into *vipalas* (0.4 seconds each) and *ghatikas* (24 minutes each).

The oldest lunar calendar still in use is Chinese. At least one estimate puts the first year of use of the traditional Chinese calendar at 2698 B.C. The Chinese government and most of the country, however, also rely on the western Gregorian calendar for day-to-day activities.

The Chinese calendar is based on 12 lunar months, with each month having either 29 or 30 days. Occasionally a special intermediate month was added to keep the lunar months in line with the solar year. The calendar runs on a 60-year cycle, at which time it begins repeating itself. Inside this big cycle, however, are 5 smaller cycles, each lasting 12 years. These 12 years are named after animals. The Chinese new year starts with the new moon that occurs at the beginning of the first lunar month, which begins when the moon is in the constellation Aquarius (roughly from the third week of January to the third week of February).

A traditional lunar calendar from the Hawaiian Islands is based on the rising of stars. Each lunar month is called a *mahina* and begins when the sun sets on the first day after the new moon phase. Months are either 29 or 30 days in length, and each day in a month is named for the appearance of the moon on that day or a period of time after a particular phase. The first lunar month begins with the rising of the Pleiades in the east (*Hakali'i* or *Hui-hui* in Hawaiian).

American Moons

American colonists brought many European traditions with them when they settled this country. Among those traditions were the names of full moons. These traditional names were often connected to religious—mostly Christian—dates. In the New World, the naming of full moons was also influenced by the traditions already established in the northeastern region by Native American tribes, mostly Algonquin. Tribes in other parts of the country often had different names for the moons, usually related to natural changes caused by the seasons.

January. The first full moon after the winter solstice, the first full moon after Yule.

> Colonial American: Winter Moon (also Yule Moon)
> Algonquin: Wolf Moon (also Old Moon)
> Lakota Sioux: Moon of Frost in the Teepee
> Cheyenne: Hoop and Stick Game Moon
> Klamath: Moon of the Little Finger's Partner
> Nez Perce: Cold Weather Moon
> Wisham: Her Cold Moon
> Kutenai: Naktasu Moon (no translation)
> Haida: Younger Moon
> Taos: Man Moon
> Laguna: Lizard Cut Moon
> Oto: The Little Young Bear Comes Down the Tree
> Zuni: Trees Broken Moon (same as July)
> San Juan: Ice Moon
> San Ildefonso: Ice Moon

February. The second full moon of the year, associated with the middle of winter.

> Colonial American: Trapper's Moon (also Snow Moon, Storm Moon)
> Algonquin: Snow Moon (also Hunger Moon)
> Lakota Sioux: Moon of the Dark-red Calves
> Cheyenne: Big Hoop and Stick Game Moon
> Tewa: Moon When the Coyotes are Frightened

Nez Perce: Budding Time
Wisham: Shoulder Moon
Kutenai: Black Bear Moon
Haida: Elder Moon
Taos: Winter Moon
Laguna: Yamuni Moon (Yamuni is an edible root)
Oto: Raccoon's Rutting Season
Zuni: No Snow on Trails Moon (same as August)
San Juan: Coyote Frighten Moon
San Ildefonso: Wind Moon

March. The third full moon of the year.

Colonial American: Fish Moon (also Worm Moon, Sap Moon, Crow Moon, Lenten Moon, Chaste Moon).
Algonquin: Worm Moon (also Crow Moon, Crust Moon, Sap Moon)
Lakota Sioux: Moon of Snow-blindness
Cheyenne: Light Snow Moon (also Dusty Moon)
Delaware: Moon when the Juice Drips From the Trees
Nez Perce: Flower Time
Wisham: The Seventh Moon (also Long Days Moon)
Kutenai: Earth Cracks Moon
Haida: Tahet Moon (Tahet is a type of salmon)
Taos: Wind Strong Moon
Laguna: Schamu Moon (Schamu is a plant)
Oto: Big Clouds Moon
Zuni: Little Sandstorm Moon (same as September)
San Juan: Lizard Moon
San Ildefonso: All Leaf Split Moon

April. The fourth full moon of the year.

Colonial American: Planter's Moon (also Easter Moon, Pink Moon, Grass Moon, Egg Moon, Seed Moon)
Algonquin: Pink Moon (also Sprouting Grass Moon, Egg Moon, Fish Moon)

Lakota Sioux: Moon of Grass Appearing
Cheyenne: Spring Moon
Illinois: Do Nothing Moon
Nez Perce: Kaket Time (Kaket is an edible root)
Wisham: The Eighth Moon
Kutenai: Deep Water Moon
Haida: Ketkakaitash Moon (no translation)
Taos: Ashes Moon
Laguna: Sticky Mud Plant Moon
Oto: Little Frogs Croak Moon
Zuni: Great Sandstorm Moon (same as October)
San Juan: Leaf Split Moon
San Ildefonso: Leaf Spread Moon

May. The fifth full moon of the year.

Colonial American: Milk Moon (also Mother's Moon, Hare Moon)
Algonquin: Flower Moon (also Corn Planting Moon, Milk Moon)
Lakota Sioux: Moon of the Shedding Ponies
Cheyenne: Time When the Horses Get Fat
Nunamiut Eskimos: Moon When the Ice Goes Out of the Rivers
Nez Perce: Kouse Bread Time
Wisham: The Ninth Moon
Kutenai: Deep Water Moon
Haida: Salmonberry Bird Moon
Taos: Corn Planting Moon
Laguna: Loam Plant Moon
Oto: To Get Ready for Plowing and Planting
Zuni: Moon No Name Moon (same as November)
San Juan: Leaf Tender Moon
San Ildefonso: Planting Moon

June. The sixth full moon of the year, also the full moon closest to the summer solstice.

Colonial American: Rose Moon (also

Stockman's Moon, Strawberry Moon, Honey
Moon, Hot Moon, Flower Moon, Dyad Moon)
Algonquin: Strawberry Moon
Lakota Sioux: Moon of Making Fat
Cheyenne: Moon When the Buffalo Bulls are Rutting
Nez Perce: Salmon Fishing Time
Wisham: Rotten Moon
Kutenai: Ripening Strawberries Moon
Haida: Berry Ripening Season Moon
Taos: Corn Tassle Appear Moon
Laguna: Corn Moon
Oto: Hoeing Corn Moon
Zuni: Turning Moon (same as December)
San Juan: Leaf Dark Moon
San Ildefonso: Flower Moon

July. The seventh full moon of the year.
Colonial American: Summer Moon (also Buck
Moon, Thunder Moon, Hay Moon, Mead Moon)
Algonquin: Buck Moon (also Thunder Moon)
Lakota Sioux: Moon when the Cherries are Ripe
Cheyennne: No name
Pima: Moon of the Giant Cactus
Haida: Killer Whale Moon
Nez Perce: Red Salmon Time
Wisham: Advance in a Body Moon
Kutenai: Ripening Service Berries Moon
Taos: Sun House Moon
Laguna: Corn Tassle Moon
Oto: Buffalo Rutting Season Moon
Zuni: Trees Broken Moon (same as January)
San Juan: Ripe Moon
San Ildefonso: Rain Moon

August. The eighth full moon of the year.
Colonial American: Dog Day's Moon (also
Woodcutter's Moon, Sturgeon Moon, Green
Corn Moon, Grain Moon, Wort Moon)

Algonquin: Sturgeon Moon (also Red Moon, Green Corn Moon)

Lakota Sioux: Moon When the Cherries Turn Black

Cheyenne: Time When the Cherries are Ripe

Cherokee: Moon of the New Ripened Corn

Nez Perce: Summer Time

Wisham: Blackberry Patches Moon

Kutenai: Berries Ripen Even in the Night Moon

Haida: Collect Food for Winter Moon

Taos: Autumn Moon

Laguna: Yamoni Moon (Yamoni is an immature ear of corn)

Oto: All the Elk Call Moon

Zuni: No Snow on Trails Moon (same as February)

San Juan: Wheat Cut Moon

San Ildefonso: Wheat Cut Moon

September. The ninth full moon of the year, also the full moon closest to the fall equinox.

Colonial American: Harvest Moon (also Fruit Moon, Dying Grass Moon, Barley Moon)

Algonquin: Harvest Moon

Lakota Sioux: Moon when the Calves Grow Hair (also Moon of the Black Calves and Moon when the Plums are Scarlet)

Cheyenne: Cool Moon

Paiute: Moon Without a Name

Nez Perce: Spawning Salmon Time

Wisham: Her Acorns Moon

Kutenai: Ripe Choke Cherries Moon

Haida: Salmon Spawning Moon

Taos: Leaf Yellow Moon

Laguna: Corn in the Milk Moon

Oto: Spider Web on the Ground at Dawn Moon

Zuni: Little Sandstorm Moon (same as March)

San Juan: All Ripe Moon

San Ildefonso: All Ripe Moon

October. The tenth full moon of the year, also the full moon after the fall equinox and the Harvest Moon.

    Colonial American: Hunter's Moon (also Blood Moon)

    Algonquin: Hunter's Moon

    Lakota Sioux: Moon of the Changing Season

    Cheyenne: Moon When Water Begins to Freeze on the Edge of the Stream

    Nez Perce: Falling Leaves Time

    Wisham: Her Leaves Moon (also Travel in Canoes Moon)

    Kutenai: Falling River Moon

    Haida: Kaganakyash Moon (no translation)

    Taos: Corn Ripe Moon

    Laguna: Ripe Corn Moon

    Oto: Deer Rutting Season Moon

    Zuni: Great Sandstorm Moon (same as April)

    San Juan: Leaf Fall Moon

    San Ildefonso: Harvest Moon

November. The eleventh full moon of the year.

    Colonial American: Beaver Moon (also Frosty Moon, Snow Moon)

    Algonquin: Beaver Moon

    Lakota Sioux: Moon of the Falling Leaves

    Cheyenne: Freezing Moon

    Nez Perce: Autumn Time

    Wisham: Her Frost Moon (also Snowy Mountains in the Morning Moon)

    Kutenai: Killing Deer Moon

    Haida: Stomach Moon

    Taos: Corn Harvest Moon

    Laguna: Autumn Moon

    Oto: Every Buck Loses His Horns Moon

    Zuni: Moon No Name Moon (same as May)

    San Juan: All Gathered Moon

    San Ildefonso: All Gathered Moon

December. The twelfth full moon of the year, also the full

moon before the winter solstice.
Colonial American: Christmas Moon (also Christ's Moon, Long Night Moon, Moon before Yule, Oak Moon)
Algonquin: Cold Moon (also Long Night's Moon)
Lakota Sioux: Moon of the Popping Trees
Cheyenne: Big Freezing Moon
Nez Perce: Heekui (no translation)
Wisham: Her Winter Houses Moon
Kutenai: Nistamu Natanik (no translation)
Haida: Kungyadikadas (no translation)
Taos: Night Moon
Laguna: Middle Winter Moon
Oto: Cold Month Moon
Zuni: Turning Moon (same as June)
San Juan: Ashes Fire Moon
San Ildefonso: Ashes Fire Moon

Blue Moons

A blue moon is usually explained as a full moon which occurs twice in the same month. This occurs every few years as the date for one full moon falls on or near the beginning of a calendar month so that the following full moon comes before the end of the month. A blue moon can also refer to a moon that appears blue, a phenomenon associated with unusual atmospheric conditions. A blue-colored moon, or one with a green color, is most likely to be seen just before sunrise or just after sunset if there is a large quantity of dust or smoke particles in the atmosphere. These particles can filter out the colors with longer wavelengths, such as red and yellow, and leave green and blue wavelengths to temporarily discolor the moon.

Time Conversion Chart

Greenwich Mean Time is also known as Universal Time (UT) when the 24-hour clock numerals are used.

UT	AST EDT: Atlantic Standard Time		EST EDT: Eastern Standard Time		CST CDT: Central Standard Time		MST MDT: Mountain Standard Time		PST PDT: Pacific Standard Time	
	AST	ADT	EST	EDT	CST	CDT	MST	MDT	PST	PDT
00	8PM	9PM	7PM	8PM	6PM	7PM	5PM	6PM	4PM	5PM
01	9	10	8	9	7	8	6	7	5	6
02	10	11	9	10	8	9	7	8	6	7
03	11	Midnight	10	11	9	10	8	9	7	8
04	Mid	1AM	11	Midnight	10	11	9	10	8	9
05	1AM	2	Mid	1AM	11	Midnight	10	11	9	10
06	2	3	1AM	2	Mid	1AM	11	Midnight	10	11
07	3	4	2	3	1AM	2	Mid	1AM	11	Midnight
08	4	5	3	4	2	3	1AM	2	Mid	1AM
09	5	6	4	5	3	4	2	3	1AM	2
10	6	7	5	6	4	5	3	4	2	3
11	7	8	6	7	5	6	4	5	3	4
12	8	9	7	8	6	7	5	6	4	5
13	9	10	8	9	7	8	6	7	5	6
14	10	11	9	10	8	9	7	8	6	7
15	11	Noon	10	11	9	10	8	9	7	8
16	Noon	1PM	11	Noon	10	11	9	10	8	9
17	1PM	2	Noon	1PM	11	Noon	10	11	9	10
18	2	3	1PM	2	Noon	1PM	11	Noon	10	11
19	3	4	2	3	1PM	2	Noon	1PM	11	Noon
20	4	5	3	4	2	3	1PM	2	Noon	1PM
21	5	6	4	5	3	4	2	3	1PM	2
22	6	7	5	6	4	5	3	4	2	3
23	7	8	6	7	5	6	4	5	3	4

Weight on the Moon

Earth Weight	Moon Weight	Earth Weight	Moon Weight	Earth Weight	Moon Weight
50 lbs	8.50 lbs	86	14.62	122	20.74
51	8.67	87	14.79	123	20.91
52	8.84	88	14.96	124	21.08
53	9.01	89	15.13	125	21.25
54	9.18	90	15.30	126	21.42
55	9.35	91	15.47	127	21.59
56	9.52	92	15.64	128	21.76
57	9.69	93	15.81	129	21.93
58	9.86	94	15.98	130	22.10
59	10.03	95	16.15	131	22.27
60	10.20	96	16.32	132	22.44
61	10.37	97	16.49	133	22.61
62	10.54	98	16.66	134	22.78
63	10.71	99	16.83	135	22.95
64	10.88	100	17.00	136	23.12
65	11.05	101	17.17	137	23.29
66	11.22	102	17.34	138	23.46
67	11.39	103	17.51	139	23.63
68	11.56	104	17.68	140	23.80
69	11.73	105	17.85	141	23.97
70	11.90	106	18.02	142	24.14
71	12.07	107	18.19	143	24.31
72	12.24	108	18.36	144	24.48
73	12.41	109	18.53	145	24.65
74	12.58	110	18.70	146	24.82
75	12.75	111	18.87	147	24.99
76	12.92	112	19.04	148	25.16
77	13.09	113	19.21	149	25.33
78	13.26	114	19.38	150	25.50
79	13.43	115	19.55	151	25.67
80	13.60	116	19.72	152	25.84
81	13.77	117	19.89	153	26.01
82	13.94	118	20.06	154	26.18
83	14.11	119	20.23	155	26.35
84	14.28	120	20.40	156	26.52
85	14.45	121	20.57	157	26.69

Earth Weight	Moon Weight	Earth Weight	Moon Weight	Earth Weight	Moon Weight
158	26.86	194	32.98	56	9.52
159	27.03	195	32.15	57	9.69
160	27.20	196	33.32	58	9.86
161	27.37	197	33.49	59	10.03
162	27.54			60	10.20
163	27.71	25 kgs	4.25 kgs	61	10.37
164	27.88	26	4.42	62	10.54
165	28.05	27	4.59	63	10.71
166	28.22	28	4.76	64	10.88
167	28.39	29	4.93	65	11.05
168	28.56	30	5.10	66	11.22
169	28.73	31	5.27	67	11.39
170	28.90	32	5.44	68	11.56
171	29.07	33	5.61	69	11.73
172	29.24	34	5.78	70	11.90
173	29.41	35	5.95	71	12.07
174	29.58	36	6.12	72	12.24
175	29.75	37	6.29	73	12.41
176	29.92	38	6.46	74	12.58
177	30.09	39	6.63	75	12.75
178	30.26	40	6.80	76	12.92
179	30.43	41	6.97	77	13.09
180	30.60	42	7.14	78	13.26
181	30.77	43	7.31	79	13.43
182	30.94	44	7.48	80	13.60
183	31.11	45	7.65	81	13.77
184	31.28	46	7.82	82	13.94
185	31.45	47	7.99	83	14.11
186	31.62	48	8.16	84	14.28
187	31.79	49	8.33	85	14.4
188	31.96	50	8.50	86	14.62
189	32.13	51	8.67	87	14.79
190	32.30	52	8.84	88	14.96
191	32.47	53	9.01	89	15.13
192	32.64	54	9.18	90	15.30
193	32.81	55	9.35		

Finding More Information

This book, even if it were twice as large, could not contain more than a fraction of the information that is known about the moon. For those fans of the moon who want to learn more, there are numerous sources to pick from. The following books are suggested for further reading, but not all of these titles are readily available. To locate a particular title, try your local public library, inter-library loan, university library, the listed publisher of the book, or a local bookstore.

Peterson Field Guide to the Stars and Planets, by Donald H. Menzel. 1975, Houghton-Mifflin.
Astronomy, A Step-by-Step Guide to the Night Sky, by Storm Dunlog. 1985, Macmillan Publishing Company.
Exploring the Moon through Binoculars and Small Telescopes, by Ernest H. Cherrington, Jr. 1984, Dover, Publications.
The Geology of the Terrestrial Planets, by Carr, Saunders, Stron, and Wilhelms. 1984, NASA. (NASA # SP-469).
The Moon, by Patrick Moore. 1981, Rand McNally & Company.
Practical Astronomy with your Calculator, by Peter Duffett-Smith. 1981, Cambridge University Press.
Pictorial Guide to the Moon, by Dinsmore Alter. 1979, Crowell.
Norton's Star Atlas, by Arthur Norton. 1973, Sky Publishing Corporation.
Welcome to Moonbase, by Ben Bova. 1987, Ballantine Books.
American Practical Navigator, by Nathaniel Bowditch. Published by the U.S. Defense Mapping Agency Hydrographic Center and other publishers.
The New Solar System, by Beatty, Chaikan, and O'Leary. 1981, Cambridge University Press.

Magazines: *Sky and Telescope, Astronomy Magazine, Odyssey Magazine* (for children)

For Kids (9-15 years):
Odyssey Magazine
The Young Astronaut Program
Boy Scout merit badges: Astronomy, Space Exploration
Girl Scout projects and badges: Space Exploration, Aerospace

Telescopes, astronomical equipment
S & S Optika, 5172 S. Broadway, Englewood, CO 80110
    303-789-1089
Celestron Telescopes, P.O. Box 3578, Torrance, CA 90503
Meade Instruments Corporation, 1675 Toronto Way, Costa
    Mesa, CA 92626
Edmund Scientific Company, Edscorp Building, Barrington,
    NJ 08007

Organizations:
Association of Lunar and Planetary Observers. Box 3AZ,
    University Park, NM 88003
American Astronomical Society. 2000 Florida Avenue NW,
    Washington, DC 20009
Astronomical League. P.O. Box 12821, Tucson, AZ 85732
Astronomical Society of the Pacific. 1290 24th Ave., San
    Francisco, CA 94122
Royal Astronomical Society of Canada. 136 Dupont Street,
    Toronto, Ontario, CANADA M5R 1V2

Computer Software:
Floppy Almanac. For IBM (256K). About $20.00 from U.S.
Naval Observatory, Nautical Almanac Office, Code FA,
Washington, DC 20390-5100.
Acecalc. For IBM PCs. $25.00 from AstroSoft, Inc., P.O.
Box 4451, Hayward, CA 94540-4451.
Apple Public Domain Astronomy Program. For Apple com-
puters (Apple II, 48K or 64K). About $10.00 from NASA
Teacher Resource Centers.

Astro-Macronomer. For Macintosh (400K). About $5.00 from GraySoft, P.O. Box 5456, Stn F, Ottawa, CANADA K2C 3J1
Astronomy Disc. For IBM (640K with Basic), Commodore 64/128 (32K), Amiga (512K). About $35.00 from Science Software, 7370 S. Jay Street, Littleton, CO 80123.
The Earth and Moon Simulator. For Apple II (48K). About $100.00 from Focus Media, Inc., P.O. Box 865, Garden City, NY 11530.
Eclipses and Phases of the Sun and Moon. For Apple II (64K). About $34.00 from Educational Images Ltd., P.O. Box 3456, West Side Station, Elmira, NY 14905.
Star Calc. For IBM (128K, MS-DOS 2.0 or higher). About $30.00 from Software City, P.O. Box 11082, Station H, Nepean, Ontario, CANADA K2H 7T8.

Moon Cycle Information

Astronomical Phenomena. Issued annually (usually 1-2 years in advance of a calendar year) by the U.S. Naval Observatory. For sale at about $3.00 from U.S. Government Bookstores in major cities, telescope stores, or from Superintendent of Documents. Also available on floppy disc (*see Computer Software, page 123*).
(*see Computer Software, page 123*)

  U.S. Government Printing Office
  Washington, D.C. 20402

The Old Farmer's Almanac. Issued annually (usually in October preceding the calendar year) by Yankee Publishing. For sale at about $2.50 at most newsstands and bookstores, or from:
  Yankee Publishing
  Main Street
  Dublin, NH 03444.

The Moon Calendar. Issued annually (usually in August pre-ceding the calendar year) by Johnson Books. For sale at about $7.00 at local bookstores or telescope stores or order from:
Johnson Books
1880 South 57th Court
Boulder, CO 80301

Metric Conversions

Fahrenheit degrees $= \dfrac{9 \text{ deg. C} + 32}{5}$

Celsius degrees $= \dfrac{5}{9} (F \text{ deg.} - 32)$

Astronomical unit $=$ 92,897,000 miles
149,504,000 kilometers
1.580214×10^{-5} light years

One inch $=$ 2.54 centimeters
One centimeter $=$ 0.3937 inches
One foot $=$ 0.3048 meters
One meter $=$ 3.2808 feet
One mile $=$ 1.6093 kilometers
One kilometer $=$ 0.6214 miles
One ounce $=$ 28.3495 grams
One gram $=$ 0.0353 ounces
One pound $=$ 0.4536 kilograms
One kilogram $=$ 2.2046 pounds

Glossary

albedo   The percentage of light reflected from the surface of a planet or moon. Albedo is determined by measuring the ratio between the light reflected and the light shining on an object, with complete reflection represented by an albedo of 1.

angular diameter   The diameter of a distant object as measured by the angle formed from a point representing an observer and the outer edges of the object.

annular eclipse   A total eclipse of the sun when the moon is farthest away in its orbit around earth and therefore having an apparent diameter which is not large enough to completely obscure the sun. During an annular eclipse, a ring of light is left uncovered around the dark circle produced by the moon.

anomalistic month   The period of time it takes the moon to go from one point of apogee (or perigee) to the next: 27.55455 days.

anorthositic rock   One of the types of rocks found on the moon at higher elevations.

aphelion   The point in a planet's orbit around the sun when it is farthest from the sun. (Opposite of perihelion.)

apogee   The point in the moon's orbit when it is farthest from the earth. (Opposite of perigee.)

apogean tide   The low tide of the month that occurs when the moon is at apogee (farthest from earth).

apolune   The point in the orbit of an object around the moon (such as a spacecraft) when it is farthest from the moon's surface.

ASA   The code used to designate the light sensitivity of photographic film. Codes are designed so that higher numbers represent the greater sensitivity and lower numbers less sensitivity.

asteroid   A body of rock or frozen liquids that is in orbit around the sun. Asteroids are sometimes considered planetoids, or small planets.

astronomical unit (A.U.)   The mean distance between the sun and the earth and used as a standard of measurement. 1 A.U. = 92,897,000 miles.

astronomy   The science dealing with objects in space.

astrophysics   A branch of astronomy, using physics to study and explain celestial objects.

axis   An imaginary line through the center of mass of an object, around which the object rotates.

azimuth   The angle measured from due south of an observer to directly under an object of interest in the sky. With the observer facing south, north is 0 degrees, east is 90 degrees, south is 180 degrees, and west is 270 degrees.

barycenter   A point marking the center of mass created when two celestial objects orbit around each other.

basalt   A type of igneous rock created from lava and found on the lunar surface in low areas. Also found on earth.

breccia   A composite rock found on the moon and formed from small pieces of different minerals. Also found on earth.

celestial equator   An imaginary extension of the earth's equator into the sky. The celestial equator is 90 degrees from each of earth's poles.

celestial mechanics   The branch of astronomy dealing with the motions and gravitational effects of celestial objects.

circumference   The linear measurement around the outside of a circle or a sphere.

colongitude   The longitude on the surface of the moon marked by the terminator, the edge of the area illuminated by the sun.

conjunction   The position of two celestial bodies when they are in line with one another as seen by an observer on earth. The new moon is also referred to as moon in conjunction with the sun. (Opposite of opposition.)

crater wall   The circular wall formed by the impact of a meteorite on the lunar surface.

crescent moon   A phase of the moon just before and after the new moon, when only a thin curved section is lighted by the sun. The last crescent moon before the new moon is sometimes called the old crescent moon, and the first crescent moon after the new moon is sometimes called the young crescent moon.

culmination The highest point a celestial body reaches in the sky as seen from earth.

dark of the moon Another name for the new moon.

Daylight Savings Time (DST) A legislated time change in some countries in which local times are moved up by one hour in the spring and back one hour in the fall ("Spring ahead, fall back").

declination The angle measured between the celestial equator and an object in the sky.

density An object's mass divided by its volume (grams/centimeter^3).

diurnal Referring to a period of one day.

DST (see Daylight Savings Time).

earthshine Reflected light from the earth, visible as a dull, red, or copper glow on the moon during lunar eclipses. Earthshine can also sometimes illuminate a young crescent moon so that the whole face of the moon can be faintly seen. This effect is often referred to as the "old moon in the new moon's arms."

eclipse The blocking of light to the moon when the earth comes between it and the sun, or the light to the earth when the moon comes between the sun and the earth.

ecliptic The imaginary plane formed by the earth's orbit around the sun or the plane formed by the apparent motion of the sun through the sky.

elliptical orbit A non-circular path formed by one body moving around another, in the shape of a "stretched" or distorted circle.

elongation The angle of a planet away from the sun or the moon from the earth as viewed from the earth.

ephemeris A publication or list that has information needed to locate a star, moon, or planet in the sky at a particular time.

equatorial tide A tide produced semi-monthly by the position of the moon over the equator.

escape velocity The speed required for an object to overcome the gravitational force of a planet or moon.

far side   The side of the moon facing away from the earth.

first quarter moon   The phase of the moon when it is 90 degrees away from a line between the sun and the earth. The angle of illumination creates a half-circle picture of the moon's surface, with the lighted half being on the right side.

full moon   The phase of the moon when it is on the opposite side of the earth from the sun and receives sunlight across its entire face, forming a circle of light. At this point, the moon is in opposition to the sun.

gibbous moon   The phase of the moon when it is getting larger after the first quarter moon phase (waxing gibbous) or smaller after the full moon but before the last quarter moon (waning gibbous).

gravity   One of the fundamental forces of nature, defined as the constant force of attraction between all objects in the universe. The gravitational force is inversely proportional to the square of the distance between the objects.

grazing occultation   An occultation by the moon of a planet or star where the path of the planet or star only intercepts the north or south limb of the moon.

Greenwich Mean Time   Time as measured from the 0 degrees longitude position of the Greenwich Observatory in England.

half-moon   The phase of the moon also known as the quarter moon, first quarter moon, or last quarter moon.

intercalation   A method of synchronizing a lunar calendar with a solar year by adding extra days or months. Extra days are known as intercalary days and extra months are known as intercalary months.

lacus   Latin for lake. An area on the surface of the moon resembling a small lake.

last quarter moon   The phase of the moon when it is 90 degrees away from a line between the sun and the earth. The angle of illumination creates a half circle of the moon's surface, with the lighted half being on the left side.

latitude   Lines of measurement around a planet or the moon, parallel to its equator. Measured in degrees, with the equator being 0 degrees and the poles 90 degrees north or south.

librations   The irregular motions of the moon in its elliptical orbit around earth which allow slightly more than half of the moon's surface to be visible over a period of time.

limb   The edge of a planet or moon which is visible.

longitude   Lines of measurement at right angles to the equator of a planet or the moon. Measured in degrees of angle from a designated line of 0 degrees. On the moon, 0 degrees longitude is at the center of the visible face, in the Sinus Medii.

lunar day   The period of time between two successive transits of the moon over the same meridian. The mean lunar day is 24.84 hours (1.035 times the mean solar day).

lunar eclipse   An eclipse created by the earth coming between the sun and the moon. Lunar eclipses always happen during the full moon phase.

lunar interval   The elapsed time between the transit of the moon over the Greenwich meridian and a local meridian.

lunar rays   Visible streaks on the surface of the moon which radiate away from some craters.

lunation   One orbit by the moon around the earth: a period of 29.53059 days. (Also called lunar month, synodic month).

lunitidal interval   The length of time between the transit of the moon and the following high or low tide.

magnitude   A numerical value indicating the brightness of an object in space.

mare **(plural:** *maria***)**   Latin for sea. An area on the surface of the moon (or Mars) that is low, dark, and formed from ancient lava flows.

mascon   An area of the moon's surface formed from dense, thick lunar material and having strong local gravitational effects.

mean   A mathematical average of a set of numbers or measurements, with the mean equaling the sum of the numbers divided by the number of units. The mean radius of the moon, for example, is the average radius figured from multiple measurements.

meridian   An imaginary line that passes directly north and south through an observer or specified location on earth. A plane extended from this line into space passes through the zenith (point above the observer).

meteor   A flash of light seen in the sky as an object enters the earth's atmosphere. A meteoroid is an object in space; a meteorite is an object after it has reached the surface of the earth, another planet, or the moon.

moon   The natural satellite of the earth or the natural satellite of any planet.

moonrise   The point in time when the upper limb of the moon is even with the earth's horizon as the moon rises in the east.

moonset   The point in time when the upper limb of the moon is even with the earth's horizon as the moon sets in the west.

nadir   An imaginary point directly under an observer on the surface of the earth, extending through the earth and into the sky. (Opposite of zenith.)

neap tide   The lowest high tide of the lunar month, occurring near the first and last quarter moon phases.

near side   The side of the moon facing the earth.

new moon   The phase of the moon when it is directly between the earth and the sun. Because sunlight is hitting only the far side of the moon, it appears dark from the earth. Reflected light from earth can sometimes make the new moon faintly visible during an eclipse.

nodes   The imaginary points at which the orbital path of the moon or other celestial body crosses the ecliptic.

nodical month   A lunar cycle measured by the moon moving from one node and back again: a period of 27.21222 days.

occultation   The movement of one celestial object behind another, such as the occultation of the star Spica by the moon.

old crescent moon   Another name for the thin crescent of the moon that is still illuminated by the sun before the moon goes completely dark at the new moon phase.

opposition   A specific point in time when a moon or planet is 180 degrees away from the sun, on the other side of the earth. The moon is full when it is in opposition. (Opposite of conjunction.)

orbital eccentricity   The degree to which an elliptical orbit is elongated. Measured by the length of the major axis divided by the distance between the foci.

palus   Latin for swamp. An area on the surface of the moon that is dark and resembles a swamp.

parallax   The angular displacement of a distant object such as a moon, planet, or star due to the movement of the earth.

partial eclipse   A lunar eclipse in which the moon only partly enters the dark, umbral shadow of the earth but is inside the secondary, penumbral shadow. Also refers to a solar eclipse when the moon does not line up completely between the earth and sun and only partly obscures the sun.

penumbra   The lighter part of a shadow that is formed by diffused light in an area around the edges of an object.

perigee   The point in the moon's orbit when it is closest to the earth. (Opposite of apogee.)

perigean tide   The high tide of the month that occurs when the moon is at perigee (closest to to earth).

perihelion   The point in a planet's orbit around the sun when it is closest to the sun. (Opposite of aphelion.)

perilune   The point in the orbit of an object (such as a space-craft) around the moon when it is closest to the moon's surface.

phases   The visible changes that the moon goes through in every lunar month, caused by the changing angle of illumination from the sun. There are four specific phases—new moon, first quarter moon, full moon, and last quarter moon—and also non-specific phase names such as waxing moon, waning moon, gibbous moon, crescent moon.

quadrature   The position of the moon or a planet when it is at right angles to the sun. The moon is in first quarter phase when it is in east quadrature to the sun and last quarter phase when it is in west quadrature.

quarter moon   The phase of the moon that can be either the first quarter moon or the last quarter moon. This phase occurs when the moon is 90 degrees away from a line between the sun and the earth. The angle of illumination creates a half circle picture of the moon's surface, with the lighted half being on the right side during first quarter moon and on the left side for last quarter moon.

radius   The linear measurement from the center of a sphere to the surface, or half of the diameter.

regression of nodes   The backwards movement of the moon's nodes relative to the direction of orbit.

revolution   The movement of one body around another in an orbit. Not to be confused with rotation.

rille   A valley or small canyon on the surface of the moon.

rotation   The spinning of a body around its own axis. Not to be confused with revolution.

Saros Cycle   A cycle of lunar months lasting 18 years and 11.3 days, the time it takes the moon, the earth, and the sun to return to the same starting position relative to each other.

satellite   An object that is in orbit around another object in space.

selenography   The science dealing with the study of the surface of the moon.

selenology   The science dealing with the study of the moon. From the Greek goddess, Selene.

sidereal month   A lunar month as measured by a return to a specific position as marked by a certain star: a period of 27.32166 days.

sinus   Latin for bay. An area on the surface of the moon resembling the bay of an ocean.

solar eclipse   An eclipse caused when the moon comes directly between the earth and the sun, temporarily blocking out sun's disk in the sky.

spring tide   The highest tides in a lunar month, occurring near new and full moon phases, when the earth, sun, and moon are aligned.

synodic month   A lunar month as measured from the point of one new moon to the next new moon: 29.53059 days.

terminator   The line formed by the edge of the illuminated portion of the moon as it moves across the surface.

tides   The cyclical movement of bodies of water or land on earth or the moon caused by the gravitational pull of the earth, moon, and sun.

transit   The point when the path of the moon, the sun, a star, or a planet takes it across the meridian.

tropical month   The time required for the moon to move from the first point of Aries and back: a period of 27.321582 days.

umbra   The darker cone-shaped core of a shadow, surrounded by a lighter penumbral shadow. Also refers to the darker center of sunspots.

waning moon   The period in the moon's monthly cycle after the full moon and before the new moon. During this period, the lighted portion of the moon's surface is decreasing.

waxing moon   The period in the moon's monthly cycle after the new moon and before the full moon. During this period, the lighted portion of the moon's surface is increasing.

young crescent moon   Another name for the thin cresent of the moon that is illuminated by the sun just after the new moon.

zenith   The imaginary point directly above an observer on earth. (Opposite of nadir.)

Index

The Moon Book is a companion text to *The Moon Calendar*, also by Kim Long. *The Moon Calendar* is an annual publication in poster format (31½" x 20½") printed in nocturnal black and lunar white, and packaged in clear, heavy plastic tubes with vinyl caps. The phases of the moon are graphically displayed for every day of the year, with information about phases, lunar eclipses, and dates of apogee and perigee. *The Moon Calendar* is also available as a gift or display card (6½" x 10½"). *The Moon Calendar* for each new year is available in August of the preceding year.

For information about ordering *The Moon Calendar* or inquiries about quantity or trade discounts, contact the publisher:

Johnson Books
1880 South 57th Court
Boulder, CO 80301
303-443-1576